So You're Going To Automate

So You're Going To Automate

An EDP Guide to Automation for Small Businesses

Jack Munyan

FIRST EDITION

 PETROCELLI / CHARTER   NEW YORK 1975

Library of Congress Cataloging in Publication Data

Munyan, Jack, 1929 –
So you're going to automate.

1. Electronic data processing – Business. I. Title.
HF5548.2M823   658'.05'4   75-11977
ISBN 0-88405-317-2

Contents

Introduction

There is one thing we learn from history, and that is, we do not learn from history. This paradoxical statement is certainly true when it comes to converting manual accounting systems to a computer. Costly errors amounting to hundreds of thousands of dollars have been made in this area of computer application, and there is every reason to believe the same errors will continue to occur.

This book highlights the problems encountered when going from a manual (or semiautomated) system to a fully automated system. It identifies the problems the accounting department of a small business faces most often during conversion, indicating the disasters that occur when they are not solved, and providing guidelines that can be followed to avoid the disasters.

The accounting activities generally considered for automation are accounts receivable, accounts payable, payroll, cost center analysis, inventory control, and general ledger. Normally these activities are already filling the minimum control and reporting requirements, but there are cogent reasons for automating them:

> To function at all, management needs information — information that measures performance accurately and is received early enough to take corrective action in time to be effective. It is the objective in automating most accounting systems to provide relevant reports that are accurate and timely.

> The tasks of middle management are continually increasing. By effectively utilizing the computer, management can relieve itself of many of the recurring, time-consuming, and routine tasks.

In this world of burgeoning complexity, the small accounting department slowly succumbs to the amount of paper that must be processed and the time required to process it. It is forced to the computer — either its own or that of a service bureau — to relieve the bottlenecks and speed its control and reporting functions. The computer's speed, flexibility, machine error-free operation, easy accessibility, and large capacity to store a tremendous amount of information affords to management a tool that can provide useful information accurately and on time. Moreover, although most accounting

vii

systems are not expected to provide answers for complex management problems, they can be programmed so that routine decision-making tasks can be handled automatically, thus allowing management to concentrate on the more creative and constructive types of decisions.

EDP systems that perform effectively do not just happen, however. They must be made to happen. But there are many pitfalls along the road to this utopia, and even the wary controller can be trapped. Many managers believe that once they get a computer, their system problems will disappear. However, very often new situations develop, bringing more complex problems. In fact, if these problems are not identified and solutions to them evolved ahead of time, they can cost the company dollars, time, customers, and employees, and even impact the "bottom line" adversely.

But the road has been well traveled, and there is experience from which to benefit. Clear understanding and proper utilization of this experience will minimize the dangers. It is the objective of this book to highlight this experience.

1

Facing the Problems Squarely

The setting is the accounting department of a small business. The accounting functions are manual (or semimanual). You are the controller. You and other members of the management team have decided for any number of reasons that your company should have a computer. You have versed yourself in both the theoretical and practical aspects of computer selection, systems design, and implementation — from books, magazines, and the manuals from sales reps of computer suppliers — so that you have a good overview of what it is all about. Each supplier stands ready to work with your company and to help in the implementation.

You are seemingly well prepared for your venture — but are you? What seems straightforward enough often develops complications. What follows will underscore the areas where you must give special attention to enable you to avoid at least some of the pitfalls that others have experienced.

If you have already decided to purchase computer equipment, Chapters 1 and 2 prove to be invaluable in reviewing your position to make certain that you have taken the necessary precautions. If, however, you are just in the stages of thinking about getting a computer, a thorough reading of the book should give you many factors to consider before taking this big step. In any case, whether you have purchased computer equipment or not, the book can still be quite helpful in providing guidelines for an overall approach to automation and in giving you elements to consider for automating six specific accounting systems: accounts receivable, accounts payable, payroll, cost center, inventory control, and general ledger.

CHECKING THE OBJECTIVES

Why automation? Be sure your objectives are clear, incisive, and have real meaning to the well-being of your company. Consider these:

Automate all the current accounting functions during the next three years.

1

Provide to the financial organization accounting systems that will reinforce the established lines of responsibility and aid in the development of cost-conscious and cost-effective management practices.

Provide useful financial reports that are accurate and timely.

The first sounds as if you are automating for the sake of automating. The second is fuzzy, and subject to wide interpretation. The third can and should be quantified. It surely has real meaning to company performance. But also consider to what extent it must be fulfilled. Don't ask for more speed in reporting than you really need, nor reports that will not be used.

INVOLVING MANAGEMENT

You need corporate management support. It will improve the attitudes of those who must interface with the computer types, and increase the involvement of the computer types in considering the needs of the company, vis-à-vis satisfying their own interests.

Token management approval is not enough. Management should participate in establishing the requirements for the computer system, and if they are to receive some of the computer output, prior to implementation they should review the design of the reports they expect to get.

PLANNING THE TOTAL SYSTEM

There are many functions to be automated — payroll, accounts receivable, accounts payable, cost center monitoring, inventory control, and perhaps general ledger. Some of the input data must be used for more than one function or "system" as you will be calling each automated function when automation is complete; and the output of some systems will be the input for others. Be sure the interrelationships are well understood in advance. The analyst must also understand them and include them in his system planning. If he doesn't, he will end up with incompatible systems that require expensive modification and rework, or alternatively impose unnecessary data preparation and computer processing in the daily computer operations.

Concurrently, you may want to plan for more sophisticated reporting — exception reports, automatic reordering, inventory requirements forecasting, etc. Be sure that the benefits to your company will justify the cost. If you try to make your venture accomplish too much in the beginning, you may end up with a cumbersome, inefficient system that is prone to operating errors and difficult to use.

Check the need to automate each function. Not everything needs to be automated. Sometimes it is easier to manually extract the data you need from the computer output and prepare the required reports. One example might be the general ledger. Be sure your analyst examines the total system and convinces you that the changes he is making in your existing systems will really lead to effective operation and cost-savings.

INITIAL ANNOUNCEMENT

You need the support of your key employees. Employees like to know how they will be affected by changes. Potentially negative attitudes can turn out to be positive if the employees are involved or, at the minimum, informed. If the objectives of the computer project are clearly stated, and the employees understand at the onset how they are likely to be affected and how they will be expected to participate, the average employee will be anxious to cooperate.

Timing is fundamental. Don't let the project advance too far before you present it to your key employees.

COMMUNICATION WITH YOUR ANALYST

Clear communication with your analyst that sharply defines the requirements will lead to smooth and successful conversion. Communication is a two-way street. It is not only necessary that you and your people take the time to define carefully what you need, but it is also incumbent on the analyst to suggest his ideas to you, calling your attention to questions and problems that he encounters. Your accounting managers must be explicit and detailed in their requirements, but the analyst must also recognize the areas where further delineation is required.

Communication also means mutual understanding. Because the controller and the analyst often speak different languages, each using his own special terms, misunderstandings may develop. The analyst has his expertise, and you have yours. Work together; your people must do the same.

Although simple to describe, these problems are difficult to solve. Here are some guidelines.

Establishment of Detailed Objectives. Develop early on with your analyst a detailed plan that shows clearly all the functions to be automated and how the resulting systems are to interface, i.e., what output data of each system will need to be used as input data, and where.

Very often, accounting systems are automated individually without an integrated systems plan. The result is that the systems will not interface efficiently. For example, some of the input data will be needed by several of the systems in different formats, and extra computer processing, or data preparation, will be required to make the necessary conversions. Sometimes the output reports will show conflicting figures, caused by different logic rules or programming techniques in systems that have been developed and programmed by different people and at different times. The net result is that some of the systems will need to be reprogrammed.

Specifications. There must be detailed system specifications. Like any other specifications, they define exactly what the systems are to do.

Specifications are often at two levels:

General specifications that define the output reports required and the input data for them, the distribution of the output reports, the origin of the input data, what has to be done to the input data, and operating schedules.

Computer-oriented specifications required for the programming.

Sometimes these are combined. However, don't permit the results you require to be clouded by the programming details. If necessary, insist on the general specifications that omit the programming details.

Who develops the specifications. The responsibility for and the preparation of the specifications are separate functions. Only the analyst can prepare them, using his expertise for information gathering, definition of the output requirements and the input data needed, report design, and development of input data preparation procedures. While the analyst must have experience in these areas, he should also understand data processing on the computer you are getting.

But who has responsibility — the analyst, his boss, or you? It is paramount that only one person have the responsibility; who he is is of less importance. He will have to make sure that the general specifications define exactly what is required or much will "fall through the cracks."

Checklist for specifications. Unlike the clerical staff in manual operations, a computer can exercise no judgment or make ad hoc decisions. All variations in the input data and what has to be done with it must be described by the specifications; special cases not included in the specifications must be eliminated from computer processing completely.

Take your time in developing the specifications. Missed specifications can be disastrous, requiring redesign, reprogramming, restructuring of the computer files, and even redoing the conversion of the data files. Some of the things that can happen are unbelievable:

Payroll checks cannot be prepared above some amount picked arbitrarily by the analyst.

On invoices, there is no provision for "cents" because historically there has been no need for this.

Programming is required every time prices are changed or discounts are modified.

The format for customer codes does not allow enough room for new customer types.

There are even more esoteric things that can go wrong:

The computation of interest or taxes is not accurate because the arithmetic operations and file formats do not permit enough precision.

The master file is designed to contain historical information (which it doesn't need), and in time becomes too big. This results in inefficient operation and late reports. Before long there will be pressure for a larger and faster computer.

Such problems can occur, and they will, unless the specifications are designed to accommodate all the variations that the accounting systems may be required to handle.

Reviewing final specifications. Insist on a written summary of the specifications for your approval before programming begins. The specifications must be clear to you, and you and your analyst should jointly approve them. You may not have taken care of everything, but at least both you and your analyst will be working from the same ground rules.

Making Sure of the End Result. The analyst is positive he has understood you and your people. Moreover, you feel confident you have clearly explained what you want. You and the analyst have jointly approved the specifications, and you turn to other matters while waiting for the programming to be completed.

But programming often takes months to complete and you will not see any of the output reports until the programs have been debugged. When you finally see the results, you may be surprised to find they are not what you thought you were getting. What you thought the specifications described is not the way they were understood by the analyst. The result is some of the statements and reports are not acceptable; moreover, the forms for them have already been ordered. Don't let it happen. Make sure you approve samples of each report before programming starts.

Documentation. Documentation is needed — for you and your people, and for the computer staff. Documentation for the latter is computer-oriented and provides details of program design, maintenance, and operation. The documentation for you and your managers shows exactly what you need to do and when, in inputting data into the computer system, and it explains what you get back and when. It is, in effect, a surrogate of the general specifications.

Your documentation should tell you only what you and your people need to know to interface with the new systems. Too much detail will drown you, and the documentation will be useless.

On the other hand, the documentation for the computer staff must be complete in all details. The details that are missing are those that hurt when the analyst or the programmer leaves the company.

Traditionally, good documentation is hard to get. If you get it, you have insurance against turnover; if you don't, hope that the personnel involved with the project remain with your company long enough to complete the project. A standard note in many computer programs left behind by the programmers is "PK." It means "Programmer Knows."

Training the Staff. Your staff needs to understand how to interface with the new systems when they begin operation. Training programs are required. Your analyst, if he has done his homework properly, has already established rapport with your staff. Be sure he prepares a training program that introduces them to the new forms to be used, the new procedures to be followed, and the new reports to be produced.

The training program must include hands-on experience; it should be completed prior to the operation of the new systems. Otherwise, the computer will be ready, the systems will be ready, but your staff will not be able to provide the input data in the correct formats on the schedule required, or analyze the output.

REINVENTING THE WHEEL

The policies and practices of most businesses are unique. They have evolved through the years, and have been responsive to the specific requirements of the business. It is traditional to design computer systems to satisfy existing practices.

However, today there are available many proprietary software packages for accounting applications. Some of the packages are available from the manufacturers of computers; many more are available from independent suppliers. There is a lot of variety, and some of the packages are quite good. Don't overlook the possibilities. It may be a lot easier to modify your own standard practices to match an existing software package than to reinvent the total system. It is embarrassing to find out after the fact about a software package that could have saved you time and money.

NOTIFYING THE "OUTSIDE WORLD"

After your system is running, it will be providing new forms of output to your customers and suppliers. Build good public relations by informing them in advance of what to expect.

Your auditors must know about the new systems and any changes in reporting should be thoroughly discussed with them. Where unions exist, they must also be involved at an early stage if there are to be any changes in job classifications, pay, and promotion opportunities.

PREPARING FOR CONVERSION AND FOLLOW-UP

The process of preparing input for new accounting systems will in general be different from the old one. There will be new input forms, new instructions for completing them, and new checks and balances. The computer will do some checking and it will produce error reports. Someone has to be trained to distinguish the various types of errors and provide data to correct them. This will be especially important during conversion of the files.

When you start conversion of a system, plan to maintain both master files, the one for your manual system and the one for your computerized system. Don't start until you are ready. Otherwise your files will quickly get out of control and you will need to start over.

Establish controls to insure that **all** of your manual records are converted. They are easy to lose. You will also have new controls on your input data and output reports. In particular, when files are being updated on the computer, check that customer records are not being updated incorrectly or dropped in error. If your controls are not working properly at the start of conversion, there can be errors in your new file that may never be detected.

And last, conversion of the manual file always takes longer than planned. Some files are incomplete; others need interpretation. Plan for these problems and include them in your schedule.

SOME PITFALLS

After you have implemented your systems, you will wish you had not let some things happen. But you can make an effort beforehand to prevent them from happening.

Tendency of Analyst to Oversell. Some analysts think that since their company has a computer, all components of all systems must be automated. For some applications, it may be as expensive to operate the automated system as the manual system. Unfortunately, many companies realize this only after spending many dollars and much time to implement a system no better than they had before conversion. Check each area of automation with your analyst to be sure automation is justified.

Ambiguity in Responsibility. You have defined your needs to the analyst, and the analyst must respond. But you may not have defined all you need, or you may have asked for output that will be expensive and not justified. The analyst has a responsibility to be alert to such possibilities and call them to your attention, but neither he nor his programmer should ever make design decisions that affect your needs without your approval.

Inflexible Design. Changes in business practice are inevitable. Many analysts do not understand this and design rigid systems. Then, when a change is needed, you will find it is too expensive to make the change. Make sure at the onset that your analyst clearly understands the character of the changes that might be needed so he can design flexibility into the system.

Unacceptable Statements and Reports. After the programs are running, you find the output being generated is incorrect or hard to read. Moreover, the output forms have already been ordered and will have to be scrapped. You are sorry you did not take our advice and insist on prototype reports before forms were ordered and programming began.

Inadequate Input Forms. You are now in operation and you find the new forms are difficult to use. They are not conducive to efficient keypunching and your people find they can't be used in certain situations. Too little thought has been given to this often neglected area.

Inadequate Programming Management. One company had four programmers implementing their accounting systems. Progress reports did not disclose any serious problems. But when the system was installed, the programs turned out to be incompatible — the programmers had neither programmed precisely to the specifications nor coordinated their work.

Inadequate Testing. Make sure you review the output reports from the final test runs. Have all combinations of input data been checked? The analyst and programmer will tell you they have thoroughly checked their programs, but it is your responsibility to determine whether or not the system is ready to be installed.

Inadequate Parallel Processing. Don't cut off your manual system too quickly. The results could be disastrous. Make sure you are satisfied with the results of the new system before you eliminate the manual system.

Overreacting to Missed Target Dates. Once target dates are missed, pressure is often exerted by management to quicken the pace and get the job done. In order to catch up, the amount of time spent on testing and documentation will often be reduced, and this in turn causes further problems. But there are no short cuts, and the best way to solve the problem of missed dates is to set realistic dates at the start and take steps to eliminate all cause for excuses. Here are some of the more common excuses:

The machine has been down too much.

The software packages we are using have bugs in them.

We haven't been getting enough checkout time.

There are too many hardware problems.

The operators are inefficient.

The keypunchers won't keypunch my program because of production priorities.

The lead programmer quit and we had to rewrite the programs.

The accounting managers made too many changes in the specifications.

I have a few more bugs to check out.

Each one of these could be legitimate. Watch for them, and work with your analyst to minimize their effect. But when dates are missed, make sure that you and your analyst reschedule the implementation realistically. It is bad enough to miss the first target date but it is inexcusable to miss the second one.

Timely Documentation. Insist on documenting the system design and the programming as the implementation progresses. It is too late to document when implementation is complete.

2

Selecting the Right Computer for Your Company

At this point you are probably asking yourself: What types of computer equipment are available for small businesses? How do I select the best computer for my company? Must I do anything before the equipment is installed? What problems will I face? Are there steps that can be taken to avoid these pitfalls? What type of computer personnel must I have? These are vital questions and must be answered correctly in order to avoid a disastrous and costly conversion. Computer selection is not an easy task and there is no magic formula that will produce the right computer for your company. However, there are factors that should be considered. Here are a few guidelines to help you in this difficult but extremely important task.

FACTORS IN COMPUTER SELECTION

There are many users who have picked their computers by subjective evaluation. Controller, don't add your name to this list. There are sound and poor factors in selecting a computer. Here are a few factors that have influenced controllers incorrectly:

I was impressed with the salesman's presentation.

We have always dealt with this company.

This manufacturer offers the best deal.

I have examined many options and to be honest, I am quite confused. I guess one computer is as good as the next.

Unfortunately, computers have been bought with such types of reasoning, but don't you be led down this path of thinking.

11

Although there are many types of computers for small businesses, there is no one computer or system that will fill the needs of every user. This is important to remember since many vendors will claim to have just the right answer for your problems. Each small business has its own peculiar characteristics and will assign different values to the desired factors under consideration. There must be a thorough examination of your company's needs followed by a thorough analysis of all the computer data that will be available to you.

Goals of Automation. One of the first and most vital steps to take before selecting a computer is to analyze and then document in detail what your company is doing. Establish general objectives and goals to be reached during automation. Make sure you know and are convinced as to why you are automating. Don't take any further steps unless you and the other members of the management team have thoroughly resolved this issue.

In addition to assisting you in establishing general and specific objectives for each of your systems to be automated, there is another good reason to document what you're currently doing. In many cases, you'll be surprised to find out what is currently happening (and not happening) in your own company. Even if you decide not to automate, inefficiencies in the manual system will be observed and changes can often be made to reduce or eliminate many of these.

Projected Use. After you know why you want the computer, you will need to develop plans for its use. It seems strange that most companies, even small ones, that demand projected budgets and detailed progress reports in their other operating areas do not apply these same techniques to the selection and purchase of a computer. How many companies have a three-, five-, or ten-year plan on computer utilization? Perhaps this is the reason companies are continually changing computers or upgrading their equipment. Remember, all of these changes cost quite a bit of money. Controller, don't select your computer before you know how you are going to use it.

Proven Equipment. There is a tendency in the computer field for a rapid acceptance of new devices and techniques before they are really proven. Don't automate for the sake of automating and don't pick equipment that has not been thoroughly tested and proven in the field. This is especially important in your initial computer purchase. You will have enough problems in just getting used to the computer without being a testing ground for a manufacturer's new piece of equipment.

Plan for Change and Expansion. Think ahead. Short-sightedness and lack of planning at this stage may limit your company's freedom in making future choices in the computer field and may even cost your company thousands of dollars for an expensive, time-consuming, and needless conversion. A piece of equipment might be acceptable for your basic accounting systems but the selection might be different if inventory and production control systems are to be automated at a later date.

Although costs and practicality are necessary in the final selection, do not consider these factors heavily in the initial planning. Establish your objectives and then design the system to meet your needs. Don't design the system for any particular type of equipment. Don't change the nature of your business to fit the computer. At the start of any automation this factor cannot be overemphasized.

SPECIFICATIONS

After you have developed a general plan for automation, specifications must be written. However, before these can be finalized, two important questions must be answered: (1) Do you want your system to be batch-oriented or do you need immediate, on-line responses? (2) Do you plan to have an in-house computer or will you rent computer time from another company?

Speed of Response Needed. In order to answer the first question, you must decide how important it is for you and your company to have immediate responses. For most small businesses this type of response is not necessary.

Batch-oriented, or batch-mode, systems. In this type of operation the user brings the input and instructions to the computer operation center, where the job is scheduled and run. The results that are obtained depend on the length of time required for the run and the priority relationship this job has to other production and request computer runs. Batch-oriented systems have many advantages and these include:

Most widely used type of system for small businesses and probably the best understood.

Operates in a productionlike atmosphere.

Cheapest to establish and to operate.

On the other hand, the disadvantages of these systems must be considered:

Slowness in getting results.

Difficulties in selecting priorities.

Unless programmed to do so, selected data from the computer files cannot be obtained easily.

Time-sharing systems. If the user has need for immediate response, the batch-oriented system will not be the best. Some form of time-sharing or on-line equipment should be considered instead. Advantages include the speed of response and that great feeling that each user believes he is the only one using the system at that time. Controller, make sure you know how much that feeling is costing you. You may not really need it and it may cost you a few wasted dollars. There is a tendency to oversell in this area. Time-sharing has its place but it is not a cure-all for all problems in data processing and is not suitable for all applications. Only applications that are extremely time sensitive should be placed on a time-sharing system. Can you wait a day for output results? If your answer is yes, and in most cases for small businesses it will be, you should go with batch-oriented systems and leave time-sharing operations to such time-critical operations as banking and airline reservations.

Owning vs. Renting. The answer to the second question will determine if you will have computer equipment in-house or if you will be renting time from another company. The latter is often done through facilities management where an outside organization takes over the computer operation function. Systems design and programming may or may not be part of this contractual arrangement. The advantages to this type of operation include:

Costs of personnel and equipment are shared.

Systems design can be more flexible since different types of computer equipment are usually available.

Your office manager can remain as an office manager. The functions required to handle this operation for the small business are very close to the manual method.

Be aware, however, of the disadvantages:

Programming and computer schedule changes cannot be made easily. A schedule has been prepared and it often takes an act of Congress to make a change.

This type of arrangement requires a great deal of trust between your management and that of the facilities management company.

The operating company may know computers but often they don't know or aren't interested in the details of running your business.

Quick programming changes and special reports are often difficult to obtain. The facilities management company is in business to make a profit and changes in normal operation require more red tape steps to be taken than if the computer operation was under your own control.

In contrast, in-house computers give you more control, flexibility, and the ability to train your people the way you want them. However, there are problems with owning your computer. If you don't believe this, ask somebody who owns one.

Writing the Specifications. Who writes the specifications is not as important as who is involved in the thinking. Make sure that you and your management team are closely involved in this extremely vital task. Specifications can be general or detailed. General specifications should be application-oriented and define what the system is expected to accomplish. Although easier to write, general specifications make evaluation of performance much more difficult. Detailed specifications are more time-consuming to complete and much more difficult to establish. They can also limit the number of computers that can meet your requirements. A practical alternative is a mixture of general and detailed specifications. This provides you with something valid to evaluate and forces you to think through the solution. Although writing specifications is an art and volumes have been written on this subject, remember to consider these points:

Write in terms of application, not machines.

Use simple laymen's terms. Describe what you need. Don't use technical terms unless you define them.

Use terms such as "desired" and avoid terms such as "mandatory."

For initial submission, do not set a definite ceiling on cost. Price should only be one of the evaluation factors.

Submitting the Specifications. After the specifications have been written, reviewed, modified, and approved, they can be submitted to a number of computer manufacturers. Each manufacturer will be able to set up a system that shows the particular advantages of his equipment. A manufacturer might also be able to offer suggestions as to how the desired results could be best produced on his type of equipment. The recommended system could accomplish the same results and yet be smaller and more efficient than you thought was possible when you designed the system yourself. There are many manufacturers from whom to choose. Here are a few guidelines to consider:

Limit the number of companies that will receive the specifications. An optimum number seems to be five.

Choose manufacturers that already have a great deal of experience in your particular application area.

Go to manufacturers with proven records. Check with other companies that have this type of equipment.

Choose manufacturers with available and proven support, both software and personnel. This is especially important for first-time computer users.

When you finally choose, stay with one vendor for all of your initial equipment needs. When you are larger and more experienced in computer operation, you may want to mix equipment and vendors.

PROBLEMS WITH COMPUTER PERSONNEL

Computer personnel can often be grouped into two main categories—operations and systems/programming. Operations personnel handle the running of the computer. Do not confuse these people with the engineers who repair and maintain equipment. These services are supplied by the manufacturer at a cost to you and are usually reserved for emergency breakdowns and normal maintenance. However, the selection and handling of operations personnel are your responsibility, and this is often a big problem for a small business. The management of a small business usually has not had experience in managing data processing and technically oriented people. And the computer people are usually technically oriented and not business-oriented. Thus, deciding on who to put in charge of the operations and management of a computer is a problem. Although large companies are faced with similar problems, they usually have established techniques for handling staff people.

Systems/programming personnel are those people who will help you to design the computer system and in some cases will also do the programming. For the purpose of this book, the term "systems analyst" will be used to include both functions of systems design and programming.

Historically, the normal approach for a small business is to combine the operations and the systems analyst function into one position, that of a data processing manager. The company may then look for an experienced man on the outside; in many cases, the small business hires a programmer. Here are two good reasons why this is not a recommended practice:

(1) The programmer usually does not have any expertise in information systems design, even though he may think he does. A programmer often confuses information systems design with technical computer systems design. He may know how to move computer data from one program to another, but in most cases he does not understand how business operates. He may think so, but are you going to let someone who doesn't have in-depth, practical, business experience run your company?

(2) In a small company the individual may also have to operate the computer. Many programmers do not like to get involved in this while others don't know how to.

If you go on the outside, look for someone who has had experience in your kind of business and has a proven record of designing and implementing successful business systems. On the other hand, some manufacturers claim they can install their equipment and prepare existing personnel with very little training.

Whichever way you decide, remember that the handling of computer personnel is not an easy task. Many businessmen feel that computer personnel are not interested in or lack an understanding of other aspects of business. Management often feels that the systems analysts are difficult to manage. One controller claims that the systems analysts have been partly responsible for this attitude. Many analysts believe that the uninitiated (and they include management) cannot understand computers. They are convinced that only analysts have the expertise to solve technical problems. This type of thinking must be corrected quickly or it will continue to give you problems.

Another controller who has had success in handling systems analysts lists two qualifications that he looks for: a proven programming competence and the ability to listen and then to follow management's instructions. The ability to communicate is another important consideration. Make sure that your computer personnel talk about what they do and what the computer does in layman's terms and not in technical computer language.

No matter where you get or train your computer personnel, arrange for their services at the time you decide to buy a computer. The analysts can then be actively involved in the planning and implementing of the various conversions. If separate operations personnel are needed, they can usually be brought in and trained at or near the delivery date of the computer equipment.

TYPES OF EQUIPMENT

After reading this far you might be asking yourself, "When am I going to find out what type of equipment is available?" Although important, resolution of the issues covered in the preceding sections is a necessary first step. Once you have done this, the selection of computer equipment is a much easier task.

Today a small business can buy a considerable amount of data-processing capability for $2,500 a month. This buys a great deal more than the former simple tabulating equipment installation. Consideration must also be given to the fact that the old-time solution to the problem of increasing flow of data was to hire more personnel, but remember, salaries have doubled or tripled in the last 20 years. The contemporary solution to this problem is to try to avoid the hiring of more personnel, and to look to the more modern and the more cost-effective alternatives. There are basically four types of computer operations that a small business could consider — the acquisition and use of a small-size regular computer, the acquisition and use of a minicomputer, the use of a service center or service bureau, and the use of a time-sharing service. Before too long there will be another alternative — microcomputers built into intelligent terminals — but these will not be discussed in this book.

Acquisition and Use of a Small-Size Regular Computer. There are many types of small-size regular computers from which to select. In any case, whichever type is selected, you'll have new problems to consider. A computer room must be selected, air-conditioning units for the computer equipment must be made available, other equipment such as computer tapes must be purchased, and the computer personnel problem must be tackled and solved. You will probably have batch-oriented systems which means that you'll need someone to operate your computer. You should also have a systems analyst to take care of your systems and programming tasks. An alternative to this is to own your computer and obtain your programming services from a management facilities organization.

Acquisition and Use of a Minicomputer. On the lower scale, there is often a fine technical line between a minicomputer and a sophisticated accounting machine. On the upper scale, there is a fine technical line between a minicomputer and a small-size regular computer. Manufacturers refer to them by different names. Make sure you know what you're getting and that you have not been sold a starting piece of computer equipment that will not satisfy your needs. Will this machine give you the type of system results you are expecting? It is vital to make sure you have thoroughly analyzed your systems requirements. Depending on the size of the minicomputer, the personnel and computer room requirements are quite similar to those of a small-size regular computer.

Use of a Service Center or Service Bureau. Many small businesses have found this type of service to be quite acceptable. If you do not have heavy or complicated systems requirements, you may be able to use any number of available software packages. If you plan to use this type of service, make sure you examine your needs thoroughly. If you are flexible, you may save yourself a few dollars in programming costs.

Some service bureaus have the type of systems people who can be of help to small businesses. Others do not. If you plan to get this type of service from a service bureau, plan to pay for it as a specific itemized item on your bill. You'll pay for it one way or another but the service bureau will have a stronger motivation to do a good job if they are billing directly for this service. If it is offered as part of the overall package, the analysts are often inexperienced. In many cases they will do more harm than good.

An alternative is for the small business to have its own systems analysts and rent computer time. Often this is practical only where there is an available supply of computers in the area.

Use of a Time-Sharing Service. Most accounting applications of a small business are not suited for time sharing. In most cases, the company has a terminal in its office and by use of communication lines can talk directly with the computer, usually a large-size one. The speed of the large computer is so quick that responses come back almost instantaneously and make you feel that you are the only one using the computer. In reality, many users are sharing the computer facilities. This service sounds great but it can be quite expensive. In addition, you will not have as much programming flexibility as with the other types. You will either use their application packages or your systems analysts will be given a number of programming restrictions. Except for isolated cases, this type of computer equipment service appears to be the least desirable for a small business.

PITFALLS TO AVOID

In review, here are some things that you should consider before you make that final selection.

(1) Don't think that computers will solve all your problems. Remember, decisions are still made by people. Don't pass the buck by saying, "The computer did it." Someone designed the logic rules to tell the computer what to do. The computer is quite dumb and will do only as it is instructed or programmed to do.

(2) Don't forget to involve your key people in selecting equipment. Your company needs good information but in order to get this information by means of a computer system, changes must take place. People must not only be motivated to change but must also be taught to change.

(3) In selecting computer equipment, remember that the decision to use a computer is just the tip of the iceberg.

(4) Design the system that best meets your needs. Don't change the nature of your business to fit the computer or the system. On the other hand, don't be so inflexible that you won't make minor

changes in your operations to utilize an existing software package.

(5) Make sure you establish very early on a good relationship between all levels of management and your computer personnel.

(6) Don't jump too fast. Go slow. Use the gradual approach. It's better to think small at first. Be careful of oversell.

(7) Document what your company is doing and design your system before you select your equipment. Good systems design can run on almost any type of equipment. Poor systems design will not run properly on the best equipment.

(8) Problems are not unique. They may not develop immediately, but they can occur at any time. Don't go to the next step until you have solved the one on which you're working. You may hope the problems will just disappear but they won't.

(9) Don't believe everything the vendor tells you. Many vendors claim to have just the right answers to your problems. This may or may not be true. Don't act alone. Exchange of information and evaluation of equipment can be obtained from your associates who are currently using the equipment, industry reviews, live demonstrations, and outside consultants.

(10) Don't make your selection on cost alone. The price of the computer should be only one of your evaluation factors. If you can justify the cost of the equipment based on your anticipated reduction of personnel, that's fine. But in many cases you will not be able to do this, and if you can't, you should know and plan for it in advance. Don't be surprised if it now costs you so many more dollars a month to be in business than it did before you had a computer.

3

A Right Approach to Automation

You, the controller, and your people are now ready to work with a systems analyst who has the responsibility for programming your accounting functions on the computer. He may be from an outside organization or he may report to a manager responsible for computer services. The manager may be you. You have nothing to worry about — except to be sure you get what you need when computer operations begin.

There is already an overall plan, and you have a pretty good idea of what has to be done. Detailed design and scheduling must still be completed. You must describe to your analyst exactly what the computer must do for you and you must plan to monitor progress and check results as the analyst proceeds to his detailed tasks. Finally, there will be implementation — programming, debugging, systems testing, and conversion. You, as controller, need an action plan to automate all of your accounting systems.

Although there are individual conversion differences in each of the six accounting systems discussed in this book, certain common elements should be included in any of your conversion plans. It seems that there have been almost as many different plans as there are controllers. Some of these have been successful but quite a few of them have been complete failures. Here is one right approach to automation.

ESTABLISHING A STEERING COMMITTEE

A steering committee consisting of yourself, one or two high ranking officers of the company, and at least one representative from the computer activity should be formed. The committee must determine what accounting systems are to be automated and in what priority. A realistic plan for implementation and a specific action plan for each system must be established. Proper resources must be allocated. Periodic reviews and evaluation must be made to ensure the company that the stated objectives are being met. To accomplish all of this, the steering committee must meet on a regular basis to monitor progress and to take necessary corrective action when needed. When schedules begin to slip, immediate action must be taken by the committee to make sure that everything is being done to correct the problem areas and to get the company back on schedule. This point cannot be emphasized enough. To deal with these problems, the steering company might have to assign additional manpower to the various projects, authorize overtime, or decide to eliminate certain portions of the system for the initial installation. You, as controller, must take an active role in these tasks. These are your systems and you will have to live with them when they are converted.

For this committee to be successful, all of the members must get involved in details. Successful systems seem to be those that have active participation by management. There is no substitution or delegation of responsibility in this extremely important area. Many reasons are given when an accounting system fails but at the top of the list is the fact that management was not totally involved from the planning stages right through to the implementation of the system. If the committee does not get involved in details, it will be nothing more than a rubber-stamp committee producing disastrous results.

As the controller, you may need an assistant to handle such tasks as preparing the user's manual written in accountant's language, training the other accountants, preparing test data, and helping in the implementation of the system. Remember, however, you can delegate your authority in these areas but you cannot delegate your responsibility.

REALISTIC PLANS FOR IMPLEMENTATION

It is vital that you have documented in detail what your company is presently doing and that you have established general specifications for the accounting systems to be automated. It is assumed that these tasks were completed before you selected your computer. If not, make sure you do these immediately.

Integrated Systems. The accounting systems must now be studied for interrelationships. Many companies run into problems by ignoring this step. They try direct conversions of what they are currently doing without giving any consideration of how the information from the various systems should tie together. The better approach is to develop systems that combine or integrate the various operations of your business. In these integrated systems the information data is recorded once as input to one system as close to the source or origin as possible so that the data can be used consistently for all required purposes. This is one way to insure that the output data of one system will be compatible with all other systems that use this data as input. This does not mean that there has to be one gigantic system or data base to handle all accounting systems. In fact, it is highly recommended that only the most sophisticated type of company use this approach. In contrast, the small business should use a step-by-step approach by designing and installing systems one at a time but always keeping in mind that each system is part of the total system. Go slowly, especially in the beginning.

The ultimate goal in this type of approach is to interrelate all major systems in a company so that the organization can be viewed as an integrated entity. Results from one system should be consistent with results from all other systems. In most companies, the applications can be classified under functioning groupings such as sales and order entry, inventory control, production planning, accounts payable, cost accounting, payroll-personnel, general accounting, and profit planning. Normally these functions can be considered as individual segments of a total system. In some cases, however, as in inventory control, it may be necessary to have several systems or subsystems such as finished goods, raw materials, tools, and work-in-process. Figure 3.1 shows an example of one grouping of functional areas.

SALES AND ORDER ENTRY	INVENTORY CONTROL	PRODUCTION PLANNING	PAYROLL/ PERSONNEL
Incoming orders	Finished goods	Methods	Payroll
Shipping orders	Raw material	Standards	Personnel statistics
Sales invoices	Tools	Scheduling	Wage statistics
Sales reports	Work-in-process	Expediting	Taxes
Backlog analysis		Resource requirements	Insurance
Accounts receivable			Fringe benefits
			Check reconciliation
ACCOUNTS PAYABLE	COST ACCOUNTING	GENERAL ACCOUNTING	PROFIT PLANNING
Purchase orders	Resource distribution	General ledger	Sales planning
A/P records	Direct labor performance	Trial balance	Cash planning
Vendor analysis	Cost center reports	Journal entries	Budgets
Check reconciliation		Investments	Forecasts
		Capital assets	
		Financial statements	

FIG. 3.1   Functional areas

These functional groupings should be examined to determine what systems should be automated and in what priority. As controller, you must play an active role in determining these priorities. Relationships must be determined at this point even though the systems are to be designed and installed separately. Figure 3.2 is an example of an input/output matrix illustrating what you need to tell the systems analyst about the interrelationships of the desired systems. Observe how accounts payable and payroll output must feed the cost center system which in turn supplies input to the general ledger system. After you complete this matrix, redo it again and then review. When you are satisfied that you have the relationships the way they should be, have your analyst develop a general flow chart to demonstrate these interrelationships.

Figure 3.3 illustrates a quite typical example of a total integrated accounting system. Although Figure 3.3 represents a large manufacturing company, it does illustrate how the various systems can tie together. Basically, the total system can be described as two main subsets of systems, with all of the individual systems supplying financial and operational information needed to operate the business. The front section pertains to the marketing functions and includes order entry, sales analysis, finished goods inventory control, and accounts receivable systems. All of these systems may have their own individual master files but can be directly interrelated by supplying certain designated date to one another. It is possible to have these four systems even more closely integrated. Order entry and inventory control could have the same product file while order entry, accounts receivable, and sales analysis could have the same customer file.

The second section pertains to the manufacturing cycle of designing a method for a product that is either a special order or is made for stock. In the cases where the product is part of the company's regular finished goods inventory, reorder points indicate that the supply must be replenished and orders are then automatically sent for the manufacturing of a designated amount, possibly an Economic Order Quantity (EOQ). The individual or special orders may require special tools and raw materials, the best methods for manufacturing must be selected (cost methods), the jobs must be scheduled (scheduling), and the cost of the jobs recorded (work-in-process). The expenses of all these jobs including labor are tied together with the company's payroll, accounts payable, cost center, and general ledger systems. All of these systems supply a great deal of operational information but these integrated systems also provide the company with a significant amount of vital financial data.

The important element of a realistic plan for implementation is to make sure that you tie together all of the major functions of your company in your initial planning. Then use a step-by-step approach of implementing your

OUTPUT

	Accounts payable	Order entry	Payroll	Inventory control	Cost center	General ledger
Accounts payable					Purchases supplies	
Order entry				Order amt. Name of customer		
Payroll					Payroll details	
Inventory control		Availability of stock	Salesmen's commissions			Inventory balance
Cost center						Expense & income details
General ledger						

FIG. 3.2   Input/output matrix

25

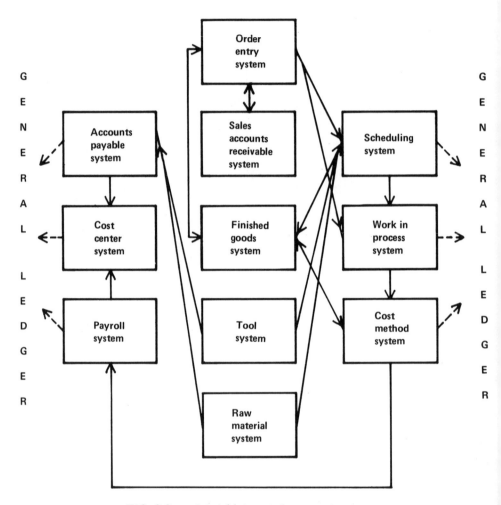

FIG. 3.3   A total integrated accounting system

systems but always keep in mind that each system is part of a total integrated system.

ESTABLISHING AN OVERALL GENERAL ACTION PLAN

Determine the order of priority for the implementation of all systems. As controller, you must establish the priorities for the accounting systems. One way to accomplish this is to separate the systems to be automated as either primary or secondary. Those systems with a primary rating could be given preliminary target dates while those in the secondary category would not have any assigned dates. An illustration of this type of action plan is shown in Figure 3.4. When the primary systems near completion, dates then could be assigned to the secondary systems. After these preliminary target dates are established for accomplishing all of the planned systems, the more detailed action schedules for each system must be developed. These must then be translated into manpower requirements and assignments.

AUTOMATING EACH OF THE SYSTEMS IN STAGES

There have been companies that have established objectives, assigned responsibility to the controller, reviewed the existing systems, and developed an overall general action plan including the assignment of priorities for

		PRELIMINARY
A.	PRIMARY SYSTEMS	TARGET DATES
	1. Payroll	December 1, 19XX
	2. Chart of accounts	January 1, 19XX
	3. Cost center accounts payable	January 1, 19XX
	4. Accounts receivable	July 1, 19XX
	5. Inventory control	December 1, 19XX
B.	SECONDARY SYSTEMS	
	1. Investments	
	2. Sales analysis	
	3. General ledger	

FIG. 3.4   Action plan

systems conversion. With this done the controller felt that he could turn the technical systems and programming assignments over to the programmer. His thinking was, "Why not? I've told him what to do. I don't know the fine points about the computer. He's the expert in that line." It doesn't take much of an imagination to determine what happened in these cases.

To avoid this type of disaster, subdivide the project schedule into six phases. These phases or steps which should be followed rigorously for each system are:

Phase 1 – System Survey

Phase 2 – System Specifications

Phase 3 – Review of Software Packages

Phase 4 – Programming

Phase 5 – System Testing

Phase 6 – Implementation and Follow-up

At the end of each of the first five phases, management should be given the opportunity to do one of the following:

(1) Give approval to the next phase. Get a firm target date for the next phase and readjust, if necessary, the other dates, **or**

(2) Make recommendations for changes. Establish a realistic target date for making these corrections. When completed, make sure that the same approval procedures stated above are followed, **or**

(3) Make a decision to cancel the project. This is always a tough decision to make but it is better to do it early before too much time and money have been expended. If this is to be done, the ideal time would be at the end of Phase 1, but no later than Phase 2.

The aforementioned technique is one recommended way for management to know what kind of job is being done. The important aspect of this benchmark check by phases is that you don't have to wait until the system is ready to be installed to discover if it was a good or poor job.

System Survey (Phase 1). Although this phase is important, it should take about 5 percent of the total time spent on the project. During this phase, a quick feasibility study is conducted to make sure it is advantageous to automate the particular accounting system. Specific project goals and objectives must be established and you, as controller, must play a leading role in this activity. What do you want to accomplish with this conversion? Make sure that you include the desired results, not just a conversion of the problems that you are currently experiencing. Although the systems analyst can assist in this area, you have the primary responsibility for establishing these specific goals and objectives.

Limitations. If certain portions are not to be automated or there are known limitations for the system, you must make these known to the analyst during this phase. You might for instance choose to develop only the payroll portion of a payroll - personnel system. The primary emphasis might be to get the payroll portion installed and then follow with the personnel section at a later date. This information must be communicated early to the analyst. Communication is a two-way street and controllers must realize and accept this responsibility of clearly and concisely telling the analyst what is wanted.

Survey report. If you have not already documented what your company is doing in the area under study, now is the time to do so. Include this information as part of your survey report. This report, which is usually completed by the systems analyst, should also include estimated costs, length of time to complete the system, and the planned benefits of the new system. There may be reduced time requirements, improved reports, more efficient handling of input, cost savings, better service, or a combination of any of these.

In some cases, you may plan on a reduction in personnel, but be careful how you handle this type of situation. If handled incorrectly, employees' fears of losing jobs could greatly decrease your chances of having a successful system design and implementation. It is very important that you avoid a very common system failure of ignoring or not involving the people or workers who will make the system go. Even if everything else in the system has been done correctly, the system must have in it what one computer expert calls "life." An effective new system cannot be planned on a technical basis alone. Key people must have an active role in the development as well as the use of the system. In order for your system to be successful, make sure that you involve and inform all key people, carefully review the documented survey recommendations, ask questions if there are parts that you don't understand, and then make your decision for go or no-go for Phase 2.

System Specifications (Phase 2). This is probably the most important phase since all subsequent work will be based on decisions made during this portion of the project. This phase should take about 40 percent of the total time spent on the automation of each system.

As controller, you have made the decision to automate, have helped in the selection of the computer equipment, and now you must determine what you want from this particular accounting system. Whether your computer

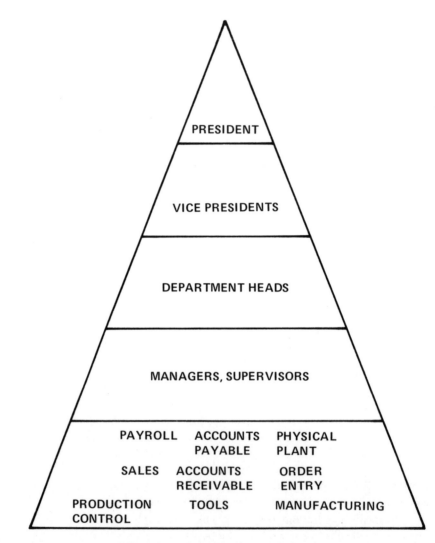

FIG. 3.5   Conceptual reporting structure

equipment is in-house or you are renting computer time, you must clearly state your requirements to the systems analyst. These fall mainly into the category of types of reports and controls that you think are necessary. The analyst should then be able to transfer these general requirements to detailed specifications.

Conceptual reporting structure. It is helpful to think in terms of a conceptual reporting structure (see Figure 3.5). Detailed reports are supplied to the lowest working level requiring computer-produced information. The next level of supervision will get a summary report of all of his area. This type of summary report will continue right up the line to the president, who will get a one-page summary of all of his direct reports. For example, if there are four direct reports to the president, the latter's report would include one or two lines of summary information for each of his staff's area. The level of detail on these reports decreases as the level of position increases. If you want this type of report structure, make sure that you explain this to your systems analyst when the general specifications are being developed.

Types of controls. You must tell the analyst what types of controls are needed for the system. Then have these controls documented by the analyst to make sure that he understands what you told him. Basically, the controls can be separated into those that deal with input, processing, or output. Examples of input control include:

(1) Validation of all inputs into the system. Make sure that the "screening" process is tight enough to allow only valid information to enter the system and be placed on the master file. Errors should be rejected and shown on an error report. Don't allow incorrect information to get on your files. It is often difficult to remove.

(2) Manual files to be completely and correctly converted.

(3) Control of number of transactions entering the system with an analysis at the end of the system of the number of good or accepted transactions plus the number of bad or rejected transactions which must equal the total of all the input transactions (Total=Good + Bad).

(4) Control of dollar totals. In some systems batch totals are used to make sure that the amounts entered (total obtained by an adding machine tape of total transactions) is equal to the amounts recorded in the computer.

Examples of processing controls include:

(1) Program to program controls to ensure that all data is passed correctly throughout the system.

(2) Proper control of master file items to make sure that no records are lost. The beginning master file count is updated by any additions and/or deletions and checked against the ending master file count.

(3)   Inclusion of "rerun" points at periodic intervals throughout the system. By doing this, possible errors can be caught without rerunning the entire system. Whenever an error occurs in the computer running of the system, it has to be corrected prior to continuing the system and thus valuable computer time can be saved.

(4)   System to system controls to make sure that all data is passed correctly from one system to another.

Here are examples of output controls:

(1)   Adequate error reports and procedures to list all errors and efficient methods of correcting these errors for reentry back into the system.

(2)   Complete and adequate documentation to cover all phases, including programming, systems, users, and operations.

(3)   All planned reports are produced when expected.

As controller, confirm that your analyst thoroughly understands what you want, but first make sure that you have thoroughly thought out what you want from the new system.

The systems analyst's design. After receiving this information, the systems analyst is now in a position to think about the fields of information, the master file, the processing rules, and the amount and method of collecting input. Consideration must also be given to timeliness of the input and output, audit and control procedures, and security. Alternative solutions to the problems must definitely be considered and analyzed by the systems analyst. When this is done, a general plan for implementation along with a general design of the system should be completed by the analyst. Then the findings should be finalized and documented for a presentation to the controller and the steering committee. Proposed input/output layouts, processing logic rules, and general flow charts should be included in this documentation. Estimated costs, manpower and time requirements should be revised, if necessary. Before any further work is started, it is imperative that as controller you approve this portion of the specifications. You cannot abrogate your responsibility by making only a cursory review. At this point, you and the systems analyst must have a clear and thorough understanding of what the system can and cannot do. Modifications may be made at this stage but they must be understood and agreed to by both sides. If there is anything that is not clear, you should insist that further details and explanations be given.

The detailed specifications. After the general design has been accepted, the analyst can now begin to develop the detailed specifications. These include detailed input/output layouts, program logic rules, and detailed flow charts. Additional data gathering and analysis are often needed to supplement the general analysis and design. You as controller must be kept informed of all of these modifications, which must then be reviewed and approved. If not, it is imperative that the differences be resolved prior to the start of the next phase of programming. At the end of Phase 2 another presentation or report should be made to the controller and the steering committee. If the specifications are not clear or are written in a language understood only by the analyst, insist that clearer explanations be given. There is no substitute for this type of detailed communication between the controller and the analyst. Once these detailed specifications are approved, the project should be frozen to changes until the programming is completed.

Review of Software Packages (Phase 3). There are many arguments, pro and con, on this subject. Here are reasons for having in-house programming:

Failure of software houses to deliver the programming packages that they promised.

Failure of packages to match company standards.

Too many problems in maintaining and updating the proprietary software packages.

Here are reasons for having proprietary software packages:

Timing needs of the company. The package might be needed in a hurry.

Budget considerations. Many people contend that software packages are cheaper than in-house programs.

Additional staff might be required to develop these systems whereas the software package approach does not require any additional people.

Whatever the company's thinking on this issue, it is recommended that at least some time be given in order to consider the following three steps as part of the decision-making process to make (develop a programming package in-house) or buy (purchase the package).

Determining the requirements. A company must determine its exact requirements and resources before it begins the evaluation process. Factors to be considered include how the task is now performed, whether it can or should be run on a computer, the type of computer selected, the location of the computer (in-house or rented computer time), and what operating facilities are necessary.

Evaluating the alternatives. The first level in this evaluating process is the make–buy decision. If you have in-house programming, you must determine in your own mind that your current programming staff is capable and experienced enough to handle this assignment. If your decision is to purchase a software package, you must then proceed through a complex and sometimes frustrating process. First you must eliminate those packages that do not offer the features that you require. The next step is to evaluate the others on the basis of their advertised features and the checked-out results of those companies currently using the package. The pricing and support aspects of the software package should also be thoroughly investigated to determine whether they meet the user's specific requirements. In addition, make sure that you understand and get a written agreement as to how maintenance and updating problems will be handled and at what cost, if any. If there are to be any modifications to the original package, make sure they are checked out and give you the required results.

Selecting the best method. There are many problems connected with software packages but as a small business it will be definitely advantageous to review existing program packages. The test seems to be that if you have a strong programming staff, do your own programming for the advantages of control and flexibility. If you are currently understaffed or are weak in the programming area, explore the possibilities of a proprietary software package.

Programming (Phase 4). During this phase, most of the work is performed by the programmers. If the programming is done in-house, detailed programming specifications must be completed, programs coded, compiled, or assembled for the computer, and then tested or "debugged." You should insist that the programmers document their programming specifications to facilitate future changes. This documentation will prove invaluable to your company in case the programmer leaves your company or is transferred to another assignment.

Preparation of a user's manual. In addition, it would be helpful if you and your staff, with assistance from your systems analyst, could prepare a user's manual. It should be written in the user's everyday language and should not contain any technical programming details. A typical manual for each system might contain: a brief description of the system; a listing of input forms with examples and instructions on how to use each form; processing rules written so that accountants can understand; a listing of all output reports with a one-page example for each report; and a copy of all procedures for this particular system.

Staff testing. While the programming is being done in-house or modifications are made by the outside programmers, you and your staff should be preparing your own test to check out the system. Remember, you don't have to know how the programmers are performing any of the technical tasks, but since you know what you want from the computer, your task is to make sure the system produces the desired results. You can do this by making up test data on the new input forms ahead of time so that you are ready for the next phase, system testing. Many of the calculations and results can be worked out ahead of time so that when the programmers deliver the output reports from your test, you can check your own predetermined results to make sure that the programs and system are operating as agreed. Keep this test deck as it can be used again, especially when future changes are made, to make sure that changes did not affect areas that had previously been working.

System Testing (Phase 5). This phase is sometimes combined with Phase 4. However, there is a natural separation, and from your standpoint, it is an important one. During Phase 4, the programmers will be checking out individual programs to make sure that these programs work. They will also conduct tests to make certain that the programs run from beginning to end. Be careful, for in some cases they may be thorough, but in others they may not. Remember, however, that the programmers' testing is most likely to be biased. They have programmed using a certain type of logic and their test data will naturally reflect their logic. Part of this difficulty could be due to a misunderstanding of what the approved specifications are. This is where you can come in with your own systems test deck. After the programmers state that all known "bugs" have been corrected and Phase 4 has been completed to the best of their ability, Phase 5, or system testing, can begin. In this phase, you should know what is going into the system, how to use the input forms, and what should be coming out of the system. From a practical standpoint, you don't have to know how the results were obtained.

Advantages. During this type of testing you can even try to foul-up the system or enter as many different combinations as possible. The goal is to simulate and test as many live conditions as possible. It is better to test at this point when there is time to react and correct problems rather than in actual production when time is usually very critical and limited. Here are a few other advantages of system testing:

(1)   The controller's area can become familiar with the input requirements, error correction procedures, and the new reports that are being generated. In order to minimize the number of pages of output reports generated, it is recommended that a miniature master file be created and used during testing. This can be done on a random or a selective basis with the number of items in the file being about 5 to 10 percent of the original total. Most combinations can be checked out without going through reams of paper.

(2)   You can assure yourself that the system is working according to agreed-upon specifications. Prior to this phase, you could only take the word of the programmers that everything was working and on schedule.

(3)   Program bugs are easier to correct during this phase than in actual production.

(4)   If there are major problems and the system installation has to be delayed, it is better to do this than to install the system with poor

and incorrect reports which will only infuriate your customers, employees, and other users of the system.

(5)   The system test deck can be used after the system has been installed and changes have been made. Users can get frustrated when one nonrelated programming change affects another area that was previously working. Many accountants cannot understand how this happens, but it does. By using the original test deck, you can make sure that the new changes are working without affecting the previously tested results. By using the miniature file, you can check your results quickly and assure yourself that the changes have been tested before they are put into actual production.

Types of system testing. Testing activities are often looked upon as questionable and unnecessary expenses. In addition, this phase has been one in which the user has not taken an active role. However, you should be totally involved with the conversion and this includes the testing of the system. There are several types of system testing and you should be familiar with the following: integrated program, or system testing; parallel systems testing; live testing; and an audit of the programmers' final systems test.

Integrated program testing is used to verify the correct interaction between programs and other systems that are either feeding or receiving data from the given system. Basically, the entire system is checked out by using data prepared by you or your assistants. The ultimate goal is to verify that all system specifications are correct and that all possible data inputs will be processed correctly. All output reports that are questioned should be resolved before installation. Although it is theoretically possible to check out all possible combinations, it does not happen this way in practice. However, it is still wise to catch as many "bugs" before production as possible. If you don't catch these errors in system testing, you'll have to solve them when you are in production.

In cases where a semiautomated system is being converted to a fully automated system, it is often possible to run both systems at the same time. Inasmuch as the old system is running concurrently with the new one, the old system could continue to be used if major problems are found in the new system. However, the specifications for the new system are often considerably different from those of the old system and this makes it extremely difficult to reconcile the differences. In addition, the amount of time to prepare input for both systems and to check different output reports is considerable, and unless significant overtime is permitted, there may not be enough time and manpower in any one day to take care of both systems. But, in those cases where it is possible (e.g., the general ledger system), it is an excellent systems test.

Live testing is checking out the system by using it as if it were operational and therefore requiring the use of live data. Live testing is usually used in the final stages of system testing when all the system components are available. An example of this would be checking out a monthly cost center system which required input from a payroll and an accounts payable system. By taking live data for the first three weeks of the month and making the assumption that the month only had three weeks of data, the entire system could be run. All reports could be checked out by taking a week before production to correct any errors found during this period of live testing. When the system is finally run in production, three weeks of data have already been checked out.

The last type of system testing is not recommended but is the minimum of what you should do. You must, at least, audit and approve the programmers' final tests. The output must be reviewed to make sure that it matches that which is expected from the given input. The problem with this type of testing is that it is usually shallow and it lacks objectivity. It tests only the programmers' understanding of the system, which is not necessarily yours.

Training of computer personnel. Another task that is included in this phase is that of training. This includes not only the users but also the operations area if you have an in-house computer. It is extremely vital that all areas being affected by the new system are thoroughly trained in the operation of the new system. It is to be hoped that all new tasks have been determined and carefully planned for; however, if it is discovered that a few tasks might have been overlooked or misunderstood, you still have time during this phase to correct the situation.

At the end of this phase you and the steering committee have a final chance to say yes or no for implementation. You must be satisfied that the system will produce the results that you want. You'll never do all the testing that is necessary, but at some point you must make your decision and proceed to the next phase.

Implementation and Follow-up (Phase 6). This last phase begins when the old system is officially converted to the new one. Here are three methods of conversion that you should know.

Immediate conversion means that system testing has been completed and Phase 5 has been approved by you and the steering committee. Old files and input are not used any more and the new system is officially installed.

Gradual conversion is where parts of a system are tested and converted over a period of time. An example of this is in a payroll conversion where there are various types of payroll, such as monthly/weekly/hourly and weekly salary.

Pilot conversions are used where there are a number of installations using the same basic system. The system at one location is tested and installed successfully before the other locations are converted.

The type of conversion selected. This will depend on the size of the system and your company. However, you must select the method of conversion ahead of time so that appropriate schedules and procedures can be set up. For example, immediate conversion requires a thorough system testing before conversion whereas gradual conversion requires that only those sections of the system currently being converted need be tested at that time.

Evaluation, a continuing process. The final step in the process is evaluation, which should continue throughout the system. Answers to the following types of questions should be obtained:

Are the objectives being met?

Are there major or minor improvements (maintenance) that should be made?

Is the documentation adequate?

You must make sure that this evaluation continues. If changes are needed, you must coordinate these changes with the systems analyst with as much lead time as possible. Often minor changes can be made easily but any major type change should pass through the same six phases with which the initial system was installed.

PROJECT SCHEDULES FOR EACH SYSTEM

The tasks needed to be accomplished for each of the aforementioned phases should be listed on a project schedule along with the individual or group responsible. In addition, target dates must be established. Care must be taken to avoid overoptimism, which can cause quite a bit of anxiety throughout the project. There is always a tendency to want to get the job done quickly. However, with such a complex operation as installing accounting systems, it is often difficult, if not impossible, to accurately foresee all of the tasks and problems that can occur in automating any accounting system. Preliminary target dates can be established at the beginning of the project but they should be reviewed and adjusted, if necessary, at the end of the benchmark phases. Only firm dates for the completion of the next phase should be established. At the end of Phase 1 (system survey), for instance, firm dates could be established for Phase 2. At the end of Phase 2 (system specifications), you should be in a better position, along with the systems analyst, to establish a firm completion date for Phases 3 and 4. Phase 2 is extremely important in establishing realistic target dates. Until all of the detailed specifications are developed and approved, you are only asking for trouble by establishing arbitrary target dates. You should never select dates unless you and your analyst know all of the tasks that have to be completed and how they are to be accomplished. Some analysts are optimistic and will accept almost any type of challenge. Remember, a missed target date is often the result of this inefficient, overly optimistic type of thinking.

These project schedules lend themselves very well to progress reports. These reports should be written and submitted by the systems analyst to you at least once a month. The report should contain a brief listing of accomplishments as they relate to the individual tasks on the project schedule; problem areas and recommended solutions; and goals and plans for accomplishing the tasks during the current review period.

When there is an indication of a missed target date, immediate action must be taken to get the system back on schedule. The first time you miss any of the target dates, you are in trouble if you don't take corrective action. You might think that this is obvious and everyone recognizes this need. But it just isn't so. Corrective action could involve additional authorized overtime, additional manpower, or higher priority for computer check-out for this system.

An example of a typical project schedule for any individual accounting system is shown in Figure 3.6. It shows certain general tasks (these could be as specific as you wish) broken down into the various phases of installing any accounting system. Manpower responsibilities and target dates are important ingredients of any project schedule and are shown after each task. Remember, these schedules are only tools, and if not properly used, are not solutions at all.

SUMMARY

This recommended approach may require more time to install systems and may even cost more because of the amount of time required of you and the other members of the management team. However, the lasting benefits of better systems and a smoother conversion should far outweigh any additional time or cost. There is an old saying that you never have time to do it right the first time but you always have time to redo it. By doing it right the first time, you will be saving time and money in the long run. The approach covered in this chapter is one way to make sure that you have a successful conversion the first time.

PHASE	TASKS	RESPONSI-BILITY	FROM	TO
SYSTEM SURVEY (Phase 1)	1-1 Prepare systems survey report	Programmer	11/1	11/15
	1-2 Review and approval of survey report	Steering Committee	11/15	11/19
SYSTEM SPECIFICATIONS (Phase 2)	2-1 Prepare report specifications	Comptroller/ Programmer	11/20	12/20
	2-2 Prepare input and processing specifications	Programmer	12/21	1/14
	2-3 Develop clerical procedures	Accounting/ Project Team	12/21	1/14
	2-4 Review and approve all specifications	Steering Committee	1/14	1/21
REVIEW OF SOFTWARE PACKAGES (Phase 3)	3-1 Make or buy decision	Steering Committee	1/22	2/12
PROGRAMMING (Phase 4)	4-1 Prepare programming specs	Programmer	2/13	2/27
	4-2 Prepare users' manual	Accounting/ Project Team	2/13	5/29
	4-3 Prepare conversion plan	Programmer	2/13	5/29
	4-4 Programming and testing	Programmers	2/13	5/29
	4-5 Develop systems test schedule	Project Team	5/1	5/25
	4-6 Approval of test schedule	Steering Committee	5/25	5/29
SYSTEM TESTING (Phase 5)	5-1 Users training	Project Team	5/30	6/15
	5-2 Computer operations training	Programmer/ Data Processing	5/30	6/15
	5-3 System testing	Project Team	5/30	6/23
	5-4 Parallel processing	Project Team	6/23	6/30
	5-5 Approval of final results	Steering Committee	6/23	6/30
IMPLEMENTATION AND FOLLOW-UP (Phase 6)	6-1 Conversion	Project Team	7/1	
	6-2 Installation	Project Team		7/2
	6-3 Follow-up and evaluation	Project Team	7/2	9/30

FIG. 3.6   A typical project schedule for any individual system

4
Guidelines to the Applications

There are no shortcuts to automation. The systems analyst responsible for designing your systems must be responsive to your needs and the needs of your staff. Make sure he listens to what you say. He should analyze your needs as you have stated them to him, and consider alternative solutions for fulfilling them. He should also be encouraged to think independently — sometimes needs can be restated to save time and money in the processing and provide more meaningful reports. Remember, however, that unilateral action must be forbidden.

In this chapter, general guidelines for each of the six accounting systems are discussed. They should help you establish effective communication with your systems analyst. You should not begin your systems design and implementation without such communication.

In Chapters 5 to 10, individual accounting systems are discussed in further detail but more from a technical standpoint. Suggestions for each system include types of reports, processing rules, master file contents, and input requirements.

If, at this stage of automation, you have decided to have an in-house computer, all of these accounting systems can be run on your small-size regular computer or a minicomputer. Whichever one you have selected, make sure there is enough computer memory to process all of your systems efficiently. If your decision was to rent computer time from another company, you will probably be running these accounting systems at a service center or service bureau.

Whether you use your own systems analysts or those from an outside computer company, these chapters provide valuable information to assist you in making sound decisions in automating. In addition, don't overlook the recommended approach in the preceding chapter, including the benchmark features used at several of the key stages of installation. Thus, you do not have to wait until the implementation stage to discover if a good or a poor job was done. At the end of each of the key stages, check the results against the stated objectives. Remember, you were involved in establishing these objectives, if you're not satisfied, don't proceed to the next phase until the problems have been corrected to your satisfaction.

Control of Accounts Receivable

Accounts receivable are the monies owed to a company by its customers for merchandise sold and for services rendered. They appear as current assets on the balance sheet and as income on the profit and loss statement.

But if the customers do not pay the monies they owe and there are not enough reserves for bad debts, the balance sheet no longer reflects the company's true worth. When bad debts are written off, the company's profit and loss statement can be significantly affected.

Equivalently, and just as serious, accounts receivable records can get out of control. When this happens, the customers do not get dunned, the accounts receivable records do not get updated, and the flow of cash into the company decreases.

Yet many controllers risk just this when they convert their accounts receivable system to a computer. What starts out as an easy conversion ends up as a morass. The controller has not clearly specified all of his needs to the analyst, the analyst has not understood the controller's problems, and no one has followed up to see if the desired results will be produced on time. When and if the programs are ready to run, the final outputs may be unacceptable to the controller. The result is that specifications must be modified, the accounts receivable customer file revised and updated, programs rewritten or modified, and the whole conversion rescheduled. Everyone has learned, but the conversion is behind schedule and the company has lost valuable time and money. If the controller has not had the foresight to continue his manual accounts receivable activity, he is in trouble.

The controller can avoid these problems, but even if he does, the task of converting accounts receivable to a computer will not be easy. If steps are not taken to avoid the problems, conversion will be costly and frustrating.

Here are some guidelines to minimize the problems you'll encounter.

ESTABLISHING MEANINGFUL OBJECTIVES

Every effective accounts receivable system should:

Provide accurate and prompt reports of payments due in all situations.

Show adequate details of goods supplied and services rendered on invoices so the customer accepts the accuracy of the invoice.

Provide for immediate flagging of delinquent accounts.

Permit the prompt processing of payments to minimize the number of statements sent out to customers who have already paid.

The automated system must be significantly faster than the manual, clearly reduce costs of routine clerical activities, and provide timely management reports.

If you can't get these results from automation, don't automate.

DETERMINING SCOPE

Before starting the design, determine how much of the system should be automated. If your problem is retail accounts receivable (many individual accounts with small individual transaction value), chances are the entire system should be automated. But even though you can justify automatic data collection devices, wait until the computer system is operating satisfactorily before you install them.

If you have a commercial accounts receivable system (a relatively small number of accounts with composite transactions and regular activity of higher value), develop an overall plan to integrate all of the related systems — order entry, sales analysis, inventory control, accounts payable, payroll, cost center analysis, and general ledger. However, install these individual systems in stages; don't try to swallow the whole ocean in one gulp.

Get one system working before you start the next. The same analyst can be used in each stage of automation and the experience that you and your analyst accumulate will be quite valuable. If you insist on installing all the systems concurrently, your chances of having a smooth conversion are slim.

DATA REQUIRED

Determine the information you need to process and analyze your accounts receivable. It is not necessarily the same data you use in your manual system. Make sure you plan for **all** the data you will need. For instance:

Is credit information needed on each customer? If so, you need to store it in the customer file.

Do you want to maintain records of a customer's past activity and payments? If so, you must provide for historical files.

You will want to compute such statistics as the number of customer accounts, the average number of transactions, the outstanding balance for each account, and the age of each unpaid item in each account. And don't forget to provide for transaction, billing, and payment dates.

If the data is ultimately needed, make sure the system can store the data and that the analyst designs his files to accommodate it — even though you may not have immediate use for it.

REPORTS YOU NEED

After you have determined the information you need, carefully identify the reports you need, describe what the reports should look like, and establish how they should be used.

Among the reports you should need is aged accounts receivable. Don't forget to tell the analyst specifically what the aging categories are, and when you want the report.

In addition to the usual accounts receivable reports, determine your need for delinquency reports, credit analysis, customer mailing lists or labels, and collection lists.

After you have defined these reports, check the data required to produce them against the items of data you have already determined you will need. You may find some omissions or even some items you do not need.

COMPATIBILITY WITH OTHER ACCOUNTING APPLICATIONS

If you do not design your accounts receivable system so the data it outputs is compatible with the other systems where the data is needed, or if it is not designed to accept data generated by the other systems, you will need to modify significantly or even reprogram the entire system after you have discovered your problem.

If, for example, customers place orders for subsequent delivery, the accounts receivable system must be compatible with order entry, the preparation of shipping documents, and invoicing. If orders are taken by sales representatives or agents working on a commission basis, data about these orders must be used for payroll and accounts payable, as well as accounts receivable. As you expand your system, some of this data will also be used for sales statistics and inventory control.

Such data interfaces make it difficult to treat accounts receivable as an independent, self-contained computer application. In fact, accounts receivable processing is frequently an integral part of more comprehensive integrated systems. One of the principal advantages in automating accounts receivable is that the same input and output data can be shared with the other accounting applications. When this is done, the amount of input data preparation and the input errors are greatly reduced. When it is not done, you can end up with costly data conversions and awkward operation; ultimately you will be reprogramming the system.

DEVELOPING THE PROCESSING RULES

For retail accounts receivable systems, the calculation of the value of individual purchases is performed at the point of sale and accepted by the customer when he signs a statement of receipt. However, make sure there is a check and balance on postings and that the system provides a cut-off date in each billing period up to which **all** customer purchases and payments are prepared and validated for processing. Check also that the current "balance forward" computation is correct.

For commercial accounts receivable systems, it is common to bill with "open item" invoices that are prepared and dispatched concurrently with the shipping documents. The pricing rules can be quite complicated. The calculations can, for example, involve a wide variety of quantity, trade, and early payment discounts. There are also the various city and state sales taxes to compute.

Be sure you thoroughly understand the specifications for pricing, and don't rubber-stamp them. A safe way to proceed is to prepare your own test data and carry out the computations according to the specifications.

Other computations performed in accounts receivable include establishing new accounts, processing payments, and debt collection. You should check each one of these.

It is common practice to change some of these computations, especially those for pricing and discounts. You must indicate to your analyst which ones are likely to change and in what ways. The analyst must build flexibility into the system design to prepare for such changes, but he must have guidelines that indicate and limit the character of the changes.

ACCESSIBILITY TO ACCOUNTS RECEIVABLE FILES

You must decide whether you will restrict your access to accounts receivable information to the regular processing cycles, or whether you require the capability to access the information at any time. One of the few advantages of manual accounts receivable systems is that the controller and other accounting personnel can interrogate any part of the accounts receivable file whenever the situation requires it. In automated systems, however, this feature can be costly, and in many cases it is not necessary. Don't require such a feature just because it is nice to have.

You must also decide how often you will update accounts receivable. Periodic updating is desirable, but not necessarily more than once each billing period.

APPROVING DATA PREPARATION SPECIFICATIONS

The design of data preparation procedures for the new system is the responsibility of the analyst. However, it is incumbent on you to understand what is required, especially where there is increased clerical effort to record the input data.

For retail accounts receivable, your principal input problem is the handling of large numbers of incoming customer payments and an even larger number of sales transaction records. Due to the large volume of these transactions, automated data collection techniques may be justified. Be sure the cost to automate is compared with the cost of time required to keypunch the input data, taking into account the errors that are inevitably introduced in manual data preparation, such as digit transpositions in customer account numbers. If the cost savings is considerable, then automation of data collection is justified. Otherwise, conversion costs could be increased well beyond the near term tangible benefits.

In all cases, confirm there are sufficient controls to assure validation of input data. If account numbers are not properly checked, sales and payment transactions can easily be recorded in the wrong account. The system design should provide a computer check for missing and invalid account numbers.

DESIGNING INPUT FORMS AND PROCEDURES

Make sure it is easy to enter sales data and customer payments on the input forms, and that they are designed for efficient keypunching when there is keypunching. In small businesses, where there is often the need to get outside keypunching assistance, it is particularly important to have good keypunching forms.

CONVERSION AND FOLLOW-UP

During conversion of the accounts receivable customer master file, plan to maintain two master files, one for your manual system and one for your computerized system. When you begin, check the total receivables in each file and resolve any differences.

When the accounts receivable files are being updated on the computer, make certain that the customer records are not lost and do periodic manual checks to assure the total receivables are correct.

Keeping Vendors Satisfied

Accounts payable are the monies owed by a company to its vendors for merchandise and services delivered on a credit basis. They appear as current liabilities on the balance sheet and usually represent purchase expenses on the profit and loss statement.

The accounts payable procedures for recording incoming vendor invoices, validating these against vouchers for accepted deliveries, scheduling payment dates, printing and dispatching checks, and reconciling canceled checks received from the bank are highly systematized and susceptible to automation. But there are many details to be accommodated in the system design and debugging has to be thorough. Conversion to the computer can take longer than planned, and during conversion it is easy for accounts payable records to get out of control. When this happens, you can lose your available discounts, and your credit rating and vendor relations can be seriously impaired.

As in the case of accounts receivable, however, you can avoid many of the problems. Here are some guidelines to help.

OBJECTIVES

Good accounts payable systems have controls to make sure you take advantage of discounts and pay your creditors the right amount and on time. Every automated payable system should:

Improve cash flow by taking the best possible advantage of discounts.

Pay bills accurately and promptly according to your payment policy.

Accurately match incoming invoices with authorized payment commitments and confirmed receipts of the goods and services ordered.

The automated system should significantly speed check processing, reduce the costs of clerical activities, and provide greater management control with more detailed and better presented prepayment schedules and post-payment records.

As controller, make sure your new automated accounts payable system will provide these features. Otherwise, don't automate.

SCOPE

Make sure you determine ahead of time how many of the purchasing procedures are to be included in the system. One major area involves invoice validation to make sure the incoming invoice corresponds to an authorized order, payment scheduling, and check dispatching.

From a strictly technical viewpoint, invoice validation can be completely computerized as long as the company concentrates management control over disbursements at the order approval stage. From an economic viewpoint, however, many small companies retain their manual invoice validation procedures because it can take as long to code an incoming invoice for computerized validation as it takes to validate it visually.

Even more important, most companies want to keep the control of the final decision on the disbursement of their cash resources in human hands. Some companies automate invoice validation and payment scheduling for all invoices below some dollar value, leaving the largest dollar invoices to be validated and scheduled manually. You, as controller, must establish how this phase of invoice payment scheduling and check dispatching will be handled.

There is also the occasional need to disburse checks immediately. Since some accounts payable systems are scheduled to run only once or twice a week, if hand checks are to be provided, manual postpayment records must be accommodated by the automated system.

Whatever decision is made on the above points, develop an overall plan enabling the accounts payable system to share a common vendor master file with any purchasing system that is planned. When and if invoice validation is fully computerized, you should have one vendor master file for the integrated purchasing and accounts payable system. But even if you plan to automate both applications at the start, get accounts payable working before you start purchasing.

DATA REQUIRED

In addition to the normal check processing information, make sure you determine what you require for vendor analysis — history of orders and performance, credit rating, and so on.

There are also those systems you may automate later — check reconciliation, projection of cash requirements, etc. And you will probably need records of your company's past payments. Remember to specify a historical data file. Now is the time to identify the data for these systems; make sure the system design provides for storing the data even though you may not have immediate use for it.

REPORTS YOU NEED

Besides the normal checks and registers that are produced, consider your need for such reports as aged accounts payable, advance schedules of payments, cash and expense projections, outstanding check lists, bank reconciliation, and vendor activity lists. Do you need a hand check register? If not, how will you handle errors on checks that have been issued?

Finally, give your analyst a schedule showing when checks must be disbursed and reports produced. But make sure the analyst designs flexibility into the system to allow for changes to the payment schedule policy. The system must also be flexible enough to inhibit the payment of any invoice until it is cleared, request the immediate payment of an individual invoice as required, and establish revised rules to reschedule payment of outstanding invoices.

COMPATIBILITY WITH OTHER SYSTEMS

Accounts payable systems must be compatible with purchasing procedures, general ledger, and cost center systems. If your general ledger system is automated, most, if not all, of the necessary accounts payable financial data can be automatically supplied to this system. If your general ledger system is not to be automated, make sure your accounts payable system does all the calculation for general ledger and outputs it on an appropriate report. Otherwise, time-consuming manual calculations will be required that could have been eliminated if the requirements had been known ahead of time.

If you require an accurate, detailed analysis of departmental expenses, be sure the input for the invoice contains a code of the type of expense and the department number to be charged. Some companies have an authorized chart of accounts file which contains all of the valid department numbers and codes. You will of course use the chart of accounts, or its equivalent, but your system must also provide for the distribution of expenses that do not have department numbers and item codes. Otherwise, you can have systems with conflicting data.

To get your expense forecast you will need to get commitments of unpaid purchase orders from your cost center system. If these commitments are needed by department, remember to include the department number and item code on purchase orders.

If your accounts payable and related systems are being designed at different times, be sure the analyst provides flexibility in the format of the cost data output files; it is always easier to program for variations in output formats than in input formats.

PROCESSING RULES

You will need to establish the rules for validating and paying invoices, allocating costs, and reconciling checks. You may want your system to check the unit prices of each item; the discounts applicable to the vendors' published price lists, which, if available, would normally be kept in the vendor master file; and exemptions on local sales taxes. The system can also check extensions and totals. If such checking is normal practice in your manual system, then you will certainly want to automate it.

CONTROL OF MANUALLY PREPARED INPUT

All computer systems are vulnerable to errors in data prepared by human hands. This is why such data must be controlled and verified. Control and verification are especially important for automated cash disbursement systems. Otherwise unauthorized and duplicate checks can easily be issued.

Compared with other accounting applications, accounts payable input is most varied and requires control of input preparation at many points. Here are some of the areas to control:

Recording new vendors or vendor changes in the vendor master file.

Recording newly authorized purchase orders.

Recording acceptance of fulfilled deliveries.

Posting incoming invoices.

Recording changes to computer-prepared payment schedules.

Recording dispatch of manually prepared checks.

Return of canceled checks by the bank.

For most small companies, automated data collection techniques are not recommended for accounts payable systems.

DESIGNING INPUT FORMS AND PROCEDURES

Make sure the input forms and procedures for recording changes to computer prepared payment schedules are easy to use. If your new system is to be an integrated purchasing/accounts payable system, redesign the purchase orders to accommodate the additional data you require and to make the form a more keypunchable document.

BEFORE AND DURING CONVERSION

In most accounts payable systems, the input documents will be sent to the computer in batches. This means the dollar total must be included on the input batch header card and must agree with the computer-calculated batch total. All errors must be recorded on an error report and subsequently corrected. Some of these errors will be clerical but others can include keypunching, programming, and machine errors. Since a printing error on a check could have serious consequences for the company's cash reserves, many controls on computer output should be included. All of these require new tasks to be performed by your accounts payable personnel. Make sure your people are properly informed, prepared, and trained ahead of time.

Before you convert, check all the combinations of input data you can identify — for example, what would happen if an invoice covers more than one purchase order, or only a portion of one purchase order; what if an incorrect vendor number is used? Prior to conversion, purge the manual vendor file of inactive vendors.

You should operate the manual and automated systems in parallel until you have developed confidence in the automated system. During conversion, check that the total for each run of the computer-produced checks equals the manual total.

Paying Employees Correctly and on Time

Automated payroll processing is probably the most common computer application in industry today. It is divisible into small segments, and computer systems of relatively modest capabilities can be utilized. This characteristic enables many organizations to install such systems.

Most people have the view that automated payroll processing is a simple application. Don't believe it. Although the same basic functions are usually included in all automated payroll systems, they can have many variations. These functions are straightforward in principle, but they can be complex in detail.

It is especially important to plan your conversion carefully. Payment of your employees — correctly and on time — is of foremost importance. Your conversion has to work. Each employee knows what his paycheck should be. When the automated system is installed, you can be sure every employee will be checking it. The problems that occur will be yours, not your analyst's.

PAYROLL POLICY AND OBJECTIVES

The payroll policy of a company is a major determinant of employee attitude. Now is a good time to check your policy. If you can improve it, do it when you convert. You can:

Simplify employee time/attendance and job reporting

Increase the frequency of payroll distribution — e.g., monthly to biweekly.

Improve benefits — e.g., insurance premiums, stock purchases.

At the same time, be sure you will get:

Faster processing of paychecks

Significant reductions in the cost of clerical activities

Greater management control in making sure the payroll is correct

Useful management reports at little or no additional cost, e.g., average earnings by job category, productive/idle time calculations, and labor accounting reports

SCOPE

Many systems combine the payroll and personnel functions, a natural combination since the same employee file is needed for both systems. However, make sure you don't jeopardize the installation of the payroll if you install the personnel portion at the same time. Better still, plan for both the payroll and personnel systems at the same time, but get payroll working first.

Another area, often overlooked and often complex, is the distribution of payroll expenses. Plan for such requirements in the beginning; otherwise you may end up with unnecessary additional programming expense and excessive computer run time.

PAYROLL SYSTEM DATA

What data is required to compute employee wages must be determined at the onset. The basic pay types include salary, hourly, exception, piece work, and incentive. Payment can be based on amount of time worked, employee output, or combinations of these, and be at standard or premium rate.

In addition, there are adjustments that can result in a pay increase or decrease; these include advances, overtime, shift differential, bonus, docking, and deductions. You must identify the data you require for all adjustments that may occur.

Other categories of data you may need are:

Employee general information about the employee — employee name, employee number, social security number, address, job code, title and position

Employee time/attendance — days worked and time at work

Employee job — the distribution of the employee's time among the various jobs to which he contributed

Deduction — the information necessary to compute statutory and voluntary deductions

Earnings — Current earnings, either in their final form or in various intermediate stages during the transformation of gross pay into net pay

"To-date" information — cumulative amounts over specified time periods, such as quarter-to-date. Certain accumulated quantities are fairly standard, e.g., gross pay and taxes withheld. Other amounts appear less frequently, e.g., net pay and various deductions.

STORAGE REQUIREMENTS

Extensive data storage files are usually needed in an automated payroll system. These files contain all the relatively constant information necessary to process the payroll — employee names, addresses, check numbers, wage rates, withholding and deduction data, etc. They may also contain year-to-date data such as year-to-date earnings and withholding. These files in aggregate are often quite similar among the various systems but the data may be divided among the files in many different ways. For some systems the data is consolidated into one large file. For others, there are many files each of which contains very specific information.

Due to the large amount of payroll data that can be kept, make sure you store only data you really need. Otherwise you may be maintaining files that can significantly increase your storage requirements and the amount of time required to process the payroll master file. Indeed, if you follow the latter practice with each of your systems, you will quickly need a larger and faster computer.

REPORTS YOU NEED

The paycheck accompanied by earnings and deductions statements and the payroll register are outputs of virtually every payroll system. Now is the time to make changes in the pay stub and the information you give to your employees. Make sure there is enough space to show each deduction. How about each type of pay? Many times requests for changes come after the first payroll is run on the new system. By then a supply of checks has been ordered and it can be months before a new stub is designed and the necessary programming changes are made.

In addition, make sure you have defined all the reports you will need to document the results of each payroll run — e.g., the deduction register, employee earnings/history record, absenteeism, idle time distribution, labor turnover, and paycheck register. There are also the periodic reports that involve the preparation and output of quarterly and annual tax reports.

It is easy to require a large number of reports. Automated payroll systems create environments prone to generating new reports and more paper. Often the same report will be produced in many different forms, or a new program will be prepared to produce a report for a one-time usage. It is not easy to eliminate a report once it's in the system. Make sure the reports you define are really needed. The paper and computer time you save will be well worth the effort.

DATA REQUIRED TO INTERFACE WITH OTHER SYSTEMS

The interface the payroll system has with the cost center system is most important. The output data from payroll that the cost center system must accommodate includes for each department a list of employees by name and number, labor distribution, and year-to-date wages. Most data for the cost center system can be prepared by sorting the data used for payroll processing. But if you want these reports, now is the time to identify the data for them; it should be stored in a single file that can be used by the payroll and cost center systems — e.g., shift worked, hours on each project, cost per hour, quantities produced.

If the company's general ledger system is to be computerized, the payroll system must also provide the required payroll data in the format acceptable to the general ledger system.

PAYROLL PROCESSING RULES

The payroll processing rules for most companies are complex. It is likely you will have several payroll schedules — weekly, bi-weekly, monthly — but you only need one payroll system; don't fall into the trap of having a payroll system for each pay schedule.

If you require a confidential payroll, you should take complete control of all data. Find out what is done with live test results, wasted reruns, and extra copies of detailed payroll reports. Because of difficulties of controlling confidentiality, some companies continue to do their confidential payroll manually or on bookkeeping machines after payroll is automated.

Make sure your analyst handles adjustments properly. This aspect is often overlooked and can cause you a lot of trouble. Current and year-to-date calculations are especially prone to programming errors; and in backing out a man's pay, check to see if the deductions are being adjusted correctly.

What type of controls do you want maintained? Often, the controls for an automated system will differ greatly from the controls you used for your manual one. Make sure you specify exactly what controls you require. You must maintain surveillance over payroll preparation, even after automation, to insure that pay is computed honestly and accurately — fictitious names can appear on the payroll just as easily after automation as before, and it is just as easy to distribute unauthorized bonuses.

INPUT

For payroll, control of input data is paramount. How are pay adjustments (increased, overtime) inputted into the system? How are they approved? How is keypunching checked? What happens to the original input? Make sure the answer to each of these questions is acceptable to you.

You also need to control time reporting. How do you record the number of hours worked for each person? How is it authorized? What do you do about sickness, time off, and overtime? How often are the inputs collected? It is in this area you can consider automated data collection techniques.

If you do not use automated data collection techniques, make sure you allow enough time for the collection, batching, and keypunching of time cards. There is still much to do — batches must be checked and corrected; programs run; error reports distributed; errors corrected; programs rerun; and checks issued, signed, and distributed to all employees. The timing and scheduling of all of these operations can be far from trivial. Don't overlook this scheduling task.

And, of course, make sure you will be collecting all the data you will require for the output reports.

INPUT FORMS AND PROCEDURES

Forms should be easy to use, but don't make them so simple that you do not get all of the required data, or that you increase the costs excessively.

Check the input procedures to make sure you do not have a paper flow bottleneck. In particular, adjustments and back-out pay transactions can be cumbersome.

PITFALLS OF CONVERSION

You cannot be too careful. The analyst will do his own system testing with his own test data, and maybe yours. But pick your own data and check the results yourself. Even with this precaution, run the manual system concurrently with the automated payroll system until you are sure it is working correctly.

If possible, install the system in stages — e.g., install the monthly payroll portion and eliminate the bugs before you attempt to install the weekly payroll. An advantage of installing the monthly portion first is that you have a longer period to correct the bugs.

When you run in parallel, the total payroll figures for the manual and the automated system should be the same. Small rounding differences in the automated system may occur due to the logic rules in the pay program. Have someone reconcile all differences and thoroughly explain them before you convert. Make sure the payroll master files contain the necessary quarter -

and year-to-date information, and other required historical data. And get a listing of the new master file to check that the necessary data has been converted correctly.

In most automated payroll systems, there will be new procedures for your payroll personnel. After the old manual system is eliminated, there may have to be a reassignment of some people due to the new automated system performing many of the functions previously done manually. Don't reassign them too quickly. Make sure that the new system is functioning properly before you make any moves.

Most important of all, be sure your payroll personnel are thoroughly trained in the importance and the details of maintaining the external controls. Include batch input controls as well as computer output controls on the number of employees, number of hours worked, and wage rates. This type of control can become very tricky due to the many types of payroll adjustments for each pay period.

Cost Center Systems

Cost center systems generally collect and compare budgeted and actual costs by department and by type. The reports generated by such systems are often called Budget Analysis and Variance Reports.

Without detailed, current departmental expense reports, it is easy to lose control of expenditures. Some departments spend a great deal more than they should; others can be well below budgeted expense because they are behind schedule. Without cost reporting it is impossible to track their performance against budget.

If you do not already have a cost center system, and you decide to implement one with your new computer, be sure you establish what you really need. You can produce information that is meaningful and useful to all levels of management, but it is much easier to grind out large quantities of data that won't be used.

There have been many cost systems that have not been effective because not enough attention was given to the needs of the managers who use them. There is a tendency to drown the managers in data. It is much better to give them data they can use.

OBJECTIVES THAT COUNT

Every good automated cost center system should provide:

An analysis and comparison of actual versus planned expenditures, grouped into meaningful categories

Timely reports on cost items relevant to the manager

Comparative historical cost reports to assist managers in establishing future budgets

SCOPE

First you need to identify the cost entries to be controlled. Just because you have a computer is no reason to report on individual cost. Group your cost types into meaningful categories for reporting, but be sure you collect and store the supporting data by cost type so you can produce it when requested.

You must also decide how much detail to require in budget submissions. It is easiest if the data corresponds one to one to the types of costs you will be collecting periodically; you can design your cost system to accumulate the budget detail into the cost categories you will control.

Variances from budget can be computed in absolute terms, in percentages, or both. Decide now what you want. Otherwise you will end up reprogramming the calculations and the formats.

When projected expenses are budgeted, they are usually computed from a schedule of standard costs for the materials and labor required. When standard costs change, you need to calculate the impact on the expenses that have been projected. If changes in unit costs occur frequently, you should reflect them in your cost variance reports and distinguish variances resulting from poor performance from those resulting from changes in the standard costs.

Detailed budget data is usually inputted once a year. The data can be submitted on a weekly, monthly, or quarterly basis; your system should be designed so the time base can be changed.

Some departments have income. You should decide at the onset if you want to show this income. However, it may be just as easy to accumulate the income separately and manually prepare reports that show income against aggregate expenses.

You also need to decide how to handle interdepartment transfers of expense, and especially whether you will distinguish it from income.

And you can do much more — provide current "burn rates" and short-term cost projections, analyze current performance against performance of previous years. etc. However, before you try to implement these more esoteric applications, first complete the basis system.

DATA YOU NEED

By the time you have established the scope of cost reporting, you should know exactly what data you will need. A cost center system obtains most of its input data from other systems. Some data is prepared specifically for the cost center system, e.g., adjustments, transfers of interdepartment expenses, and journal entries.

OUTPUT REPORTS

Every cost center system should provide a summary by department of all its costs. There should be one line for each cost category that is controlled. The report should at least include current month's charges, year-to-date charges, annual budget, and variance from budget.

Reports can be weekly, biweekly, monthly, or perhaps even less frequently. You will want prompt reporting, so devise a system of data collection with your analyst to assure the necessary input data is obtained in the shortest possible time.

You also need to determine who gets reports and what reports. The president might receive a one-page summary of cost activity for each of his vice presidents, the vice presidents would receive summary reports for their respective areas of responsibility, and the department managers would receive more specialized reports with still greater detail. If budgets have been properly developed, management can quickly determine what departments are under control and what departments require corrective action.

You may decide to use exception reports which indicate only those departments and expense items that need corrective action. But if you do decide on this instead of full reporting, show both the favorable and unfavorable exceptions.

You will probably decide to maintain historical data for year-to-year comparisons, especially if your company is affected by seasonal factors. Otherwise, you will want a comparison of current monthly figures with those of preceding months.

Take time in designing your summary reports. Reports that contain year-to-year comparisons as well as data of preceding months require special attention to make them easy to read. Another tip – use a "dash" sign, not "cr" to show negative figures; users of these reports are generally not accountants.

Finally, you will need to decide how many copies of each report are required, and how the reports are to be sequenced for distribution. Some of the reports may reflect data from the confidential payroll, and, if so, you may need to establish special rules for this distribution.

PROCESSING RULES

The processing rules for cost center systems are relatively simple and straightforward. You do not need to worry about computational accuracy (rounding errors) as you do in accounts payable, and there are no complicated tax calculations. If you forecast expenses, you will need to devise a forecasting formula, but even this should not be complicated and it will depend largely on the data supplied by the managers.

The primary problems lie in specifying the rules for year-end closing and for report formatting. All of these problems are relatively easy to solve but they are often overlooked.

Here is a checklist:

What totals, and what summary figures at the end of each report?

What calculations on each report for fringe benefits and overhead costs?

Processing rules for monthly and yearly payroll accruals?

What about purchase orders not paid at year-end?

For the sophisticated system, what rules for projecting expenses to year-end, with variances from budgets indicated.

Some of the processing rules have to do with calculations at year-end. Specify the rules for year-end calculations at the onset; if you wait until the first year-end closing, you may not have enough time to do the programming necessary to get your year-end reports when you need them.

IMPLEMENTATION OF THE SYSTEM

Installation of the cost center system is straight forward, whether or not you already have a manual system. While you do not want the implementation to drag on, there are no crises that will erupt if you do not meet the schedule you established for it. Of course, you cannot begin until the payroll and accounts payable systems are working since they are principal sources of input data.

You should take all the time you need to do thorough testing; some of this can be "live" testing. After the system appears to be error-free, begin distribution of the reports to your managers before you discontinue distribution of any manual reports you have. The managers will be very responsive in advising you of problems and discrepancies in the reports, and some of these undoubtedly will identify system errors. Finally you will be ready to cut off the manual system.

The best time to install the cost system is the beginning of the fiscal year. There are no year-to-date figures to worry about and you can be working with new budget figures.

Keeping Inventory under Control

Inventory is the quantity of goods and materials on hand. The total cost of items in inventory appears as a current asset on the balance sheet; the cost of inventory used appears in the company's profit and loss statement.

If inventory records are not accurate, it is easy to accumulate the wrong inventory items and deplete the fast-moving ones. When this happens, customers get dissatisfied and inventory doesn't get sold.

In spite of such dangers, many controllers risk losing control of their inventory when they convert their inventory system to an automated one. What starts out as an easy system to convert ends up as a disaster. With proper planning and control, you can minimize your problems, but even so, your conversion may not be easy.

Here are some guidelines that should help.

OBJECTIVES

Every good automated inventory system should be able to:

Maintain optimum inventory levels

Minimize capital investment in inventory while providing maximum customer service

Detect trends in demand quickly so that production can be matched to requirements

Identify obsolete inventories

SCOPE

Some companies can justify completely automated inventory control with sophisticated inventory forecasting; others only need a system that automatically updates and reports inventory status, with the remaining requirements accomplished manually. Moreover, it may be enough to apply the automated system to the active items, or those susceptible to forecasting techniques.

If you need only inventory status updating and reporting, don't let your analyst sell you sophisticated extensions that may not work. Exception reporting, for example, works well if you know how to set the limits for each item, but otherwise, you will spend much of your time checking on the status of items that are not being reported. Automated forecasting works well when demand patterns are relatively smooth and there is feedback from sales, but in other cases you may get completely unrealistic forecasts.

On the other hand, if you have a large number of inventory items and high transaction activity, you will undoubtedly need extensive automated capabilities. Make sure you analyze your inventory correctly and remember that different forecasting techniques might have to be applied to your various classes of inventory. The selection of the proper techniques could easily save your company considerable money, e.g., through a 10 percent (or more) reduction in inventory. A good automated inventory system can do this.

You must also consider the interfaces with your other systems. For example, in many applications, inventory control and order entry should be integrated into a single system; at the least the same product file should be used. If the rules get too complicated, however, install one system at a time and design for future integration. For most small companies such integration is not necessary and can result in many frustrating months of conversion without appreciable benefits.

DATA FOR YOUR INVENTORY CONTROL SYSTEM

If you already have an inventory system, you are probably keeping your data on cards of some kind. It is easy to decide this is all the data you need. But if you are planning a more elaborate system, in all likelihood you will need more data. For a relatively sophisticated system, you will need for each item of inventory:

Reorder Point (ROP) — the quantity that signals the need to replenish it

Economic Order Quantity (EOQ) — the most economical amount that should be ordered or manufactured

Safety Stock — the amount of "cushion" that is carried in case actual demands exceed expected usage

There is also more mundane data — how often are unit costs to be checked, what are the lead times required to replenish each item, etc.

This data is the minimum needed to implement an automatic recording system. For such a system, the quantity of each item is based strictly on the ROP, EOQ, and Safety Stock for the item. But demand changes, and you will need to provide in your system, manual or automatic, a way to respond to changes in demand. Any such scheme will depend on sales forecasts, projections from historical data, or a combination of these. If you decide to automate the projection of demand as a part of your system, you need to decide at the onset what, if any, historical data you need and start accumulating it at the earliest possible date. Don't wait until your system is nearly implemented; you may need a year or more of historical data to use the system.

All EOQ formulas require cost data — the cost of each item, the cost of carrying the inventory, and for manufacturing concerns, the cost to set up for production an item not in production. You will also need data for the various types of inventory adjustment transactions — items returned, scrapped, and transferred to work-in-process, and adjustments for physical inventory.

Finally, if you maintain several warehouses, you may want individual inventory reports for each warehouse, in addition to an aggregate report. If such reports are needed you will need to keep the same data for each warehouse.

INVENTORY REPORTS

The usual inventory status report itemizes items in inventory, showing name, identification number, quantity on hand, and perhaps dollar value and amount shipped on a current and year-to-date basis. Here are some more reports to consider:

An "ABC report" that classifies inventory items into, perhaps, three classes, providing usage and dollar value data for each class

A dead stock report showing items having unacceptable activity for a long period of time, perhaps two years

Items to be reordered or manufactured

Items out of stock

Monthly summary of inventory evaluation

If automatic inventory forecasting is implemented, a report showing actual experience versus forecast

INTERFACES WITH OTHER SYSTEMS

When an incoming order is received, inventory must be checked to see if the items ordered are in stock. For items not in stock, back order data must be generated for transmission to the customer and for whatever management action may be necessary. This means the order entry system must have access to the inventory status file, and the design of the file must be compatible with the order entry system.

At the other end are the manufacturing systems, manual or automated. If you have standard costs for your manufactured products, you will have to input these costs into your inventory control system for each item of inventory. The inventory system must accommodate additions to inventory from work-in-process and changes in standard cost data. Whether this data is maintained manually or automatically, the data must be formatted and organized so that inventory status can be updated by the inventory system.

Finally, you will need a monthly evaluation of inventory for your general ledger system, whether the latter is manual or automated.

PROCESSING RULES

It is easy to be inactive in establishing the processing rules, especially if your inventory system has automatic forecasting that embodies relatively sophisticated techniques. Many of the formulas require a technical formulation, and some controllers take them for granted. But you only need to assure yourself that the formulas are being correctly selected for your application and that you know what they are supposed to do. You don't need to develop them yourself.

Processing rules that you should develop with your analyst are:

Handling of adjustments

Handling of partial orders and orders that can only be partially filled

Handling of items in transit between warehouses

Handling of differences between cycle counting and inventory status records, or the equivalent, if you have periodic full inventory instead of cycle counting

Validation of inventory identification numbers, especially important in automatic reorder systems

You also need to decide on the frequency of processing:

Calculation of EOQ's, ROP's, and Safety Stock

Analysis of demand

Inventory status report

Changes in unit costs

You must understand how the processing rules have been implemented. If errors are made, you can easily end up with both overstock and outages.

PLANNING THE CONVERSION

As with other systems, take your time in converting. Test the daily, weekly, monthly, and year-end run. Make especially sure that there is adequate testing on the calculations for reorder points, economic order quantities, safety stock, and expected usage. You also need to compare your manual inventory status file with your newly created computer master file of inventory items. To facilitate the comparison, a means to compute subtotals for various classes of inventory should be built into the computerized system.

Finally, it is just as important to conduct a training program for the inventory control people as it is for the employees involved in the other systems. The inventory control people must understand how the new system functions and how to analyze the valuation reports to detect improper buildups and declines of inventory. You can make them feel like a part of the team if you involve them in testing the inventory system, e.g., by asking them to compare the data of the manual and automated systems while you are running the two in parallel.

General Ledger Systems

The general ledger is the historical record of all the financial transactions of a company. Its data is obtained from many areas — sales, accounts receivable, accounts payable, payroll, inventory control, etc. Some companies subdivide the general ledger into subsidiary ledgers for each of their accountable divisions.

The general ledger as well as its subsidiary ledgers are used as data bases from which a variety of periodic reports are produced for management, the auditors, and the stockholders, to enable them to measure performance against previously stated goals. Compared with most other data-processing applications, the volume of transactions entered into the general ledger system is generally small. Because of this it may not be advantageous to automate the general ledger. However, it is important that the various automated systems generating data for the general ledger provide it in a form in which it can easily be used. If all other systems are automated, automation of the general ledger permits it to be maintained up-to-date almost fully automatically as a secondary result of the operation of the other systems.

Whether or not you automate your general ledger, you must design your automated systems to generate data in a form in which it can easily be used for general ledger. If you do decide to automate, here are some guidelines to follow.

OBJECTIVES

General ledger procedures provide the data necessary to monitor the financial operations of a company and its divisions. Every general ledger system, whether manual or automated, should provide:

Continuous and accurate monitoring of cash position

Identification of principal growth points and profit earners, as well as principal contributors to loss

Immediate recognition of substantial deviations in management goals

When general ledger routines are automated, you should get the results faster and more frequently, and the cost of routine clerical activities should be reduced. If you can't get these results from automation, don't automate; recognize also that you may not be able to save enough costs or speed the reporting enough to justify automation.

SCOPE

General ledger systems usually consist of two principal parts:

(1)   The posting of all financial transactions to a data base of general and subsidiary ledger accounts.

(2) The periodic analysis of the data base to produce trial balances, profit and loss statements, balance sheets, and other financial reports required by management, the controller, and the auditors.

Generally, if you automate either part, you should automate the entire system.

Many general ledger systems produce budget variance analyses, cost variance analyses, departmental cost comparisons, and historical performance comparisons. However, all of these can be included in your cost center system. It is good practice to install your other accounting systems first, and then decide whether or not to automate the general ledger portion.

DATA REQUIRED

The general ledger may have a few dozen to several hundred distinct accounts, depending on the size and complexity of the business. There can be an equal number of accounts in each of the subsidiary ledgers. The accounts are categorized generally as either balance sheet or operating accounts. Most transactions affect both an operating account and several balance sheet accounts.

The balance sheet accounts show the tangible assets and liabilities. You must specify to the analyst what data is on each balance sheet account record, e.g., the ledger account number, description, opening and closing balances, and total debits and credits. You should decide at the same time whether you want entries on the opening balance and cumulative debits and credits for two or more distinct accounting cycles.

The operating accounts give a historical cumulative record of the company's income and expenses during one or more accounting cycles. You must identify the operating accounts you wish to maintain and exactly what items of data are to be cumulated in them.

REPORTS

The reports you want are obvious — the trial balances, the balance sheet, comparative balance sheets, profit and loss statements, and income and expense statements. If a company is affected by seasonal factors, monthly figures should be compared with those of the same month in preceding years and current year-to-date figures with those of the corresponding portion of preceding years. Otherwise, it should be sufficient to compare current monthly figures with those of preceding months. In either case, you will want to highlight the variance from performance in absolute terms and percentages.

There are also the analysis and trend type reports — year-to-year comparisons of current assets and quick assets, the sales and profit trends, ratios of current assets to current liabilities and shareholders' equity to retained earnings, the relation of total cash to estimated payroll, and the sum of cash and current receivables to the sum of payroll and accounts payable.

Undoubtedly the analyst is aware of most of these types of reports and has his own ideas about them. But these reports are fundamental to the management of your company, and unless you work with your analyst to insure that the reports show the right information in the right way, you will never complete your conversion from the manual system.

INTERFACES WITH OTHER ACCOUNTING APPLICATIONS

In general, only transaction summaries will be posted to the general and subsidiary ledger accounts, and the number of input records to your general ledger system will be small. Thus, the detail transactions will be processed in the other accounting systems, and the general ledger accounts will be used only for recording the sum of all transactions of a specific type.

You should make sure the other accounting systems do the transaction processing needed for general ledger. All the accounting applications are involved — accounts receivable, accounts payable, inventory control, job costing, cost center analysis, and payroll.

PROCESSING RULES

The processing rules are relatively simple. However, you should make sure there are checks' and balances for the various types of postings to minimize errors. You also need to decide how often you will post and to what accounts if there are to be postings to multiple ledger accounts.

DATA PREPARATION

If the other accounting systems have been computerized and the output data from them is compatible with the input requirements of the general ledger system, then the transactions and transaction summaries can be posted automatically. Otherwise, the data will have to be manually prepared, keypunched, and verified. This data should include indicators to specify to what ledger accounts the transaction and transaction summary data are to be posted.

CONVERSION

General ledger is one system for which the manual and automated versions can be easily run in parallel. There is no need to hurry, and you should take all the time necessary to reconcile differences that occur in the outputs of the two systems. Furthermore, much of the system testing can be accomplished during the conversion as you can run both systems with live and complete data.

Just because general ledger is relatively simple in comparison to the other systems, don't minimize the training you give your personnel. If timeliness of reports is one of the benefits, be sure your people understand them and can analyze them rapidly. No benefits are achieved if the reports sit on somebody's desk for several days.

5

Automating Accounts Receivable

All accounts receivable systems (manual or automated) contain the following basic elements:

Establishing customer accounts after the customer's credit has been screened by the credit department

Recording all credit transactions affecting the accounts

Billing customers for payments due

Processing incoming customer payments

Periodically preparing trial balances and aging accounts

Following up late-paying and delinquent accounts

There are wide variations, however, between industries and companies both in performing these basic procedures and in the number and character of additional procedures or functions that may be performed at the same time. The major variations occur between retail accounts receivable systems, where the customers are individuals, and commercial accounts receivable systems, where the customers are usually other companies.

In **retail accounts receivable** systems, the number of individual accounts will normally be large, the value of individual transactions small, and the rate of activity highly variable both between different accounts and for the same account at different seasons. The primary problem to be solved, with or without the computer, is one of data handling versus computation. Apart from the special cases of mail-order houses and utilities, the customers will already have collected the goods or services when the transactions are recorded for the retail accounts receivable system. Such systems can thus be designed as largely independent self-contained systems for processing purely

71

financial transactions. A typical retail accounts receivable system is illustrated in Figure 5.1.

Commercial accounts receivable systems, on the other hand, are characterized by a smaller number of accounts with a more regular activity of much higher value and often composite transactions. When the nature of the business is such that customers place orders for subsequent delivery by the supplier himself, the latter's accounts receivable system must frequently interface with the systems for order entry, preparation of shipping documents, invoicing, and therefore with sales statistics and inventory control.

When orders are collected by sales representatives or agents working on a commission basis, outputs from the invoicing and, in some cases, the incoming payments processing phases will be required by the payroll and the accounts payable systems.

These various data interfaces make it more difficult to treat commercial (vis-a-vis retail) accounts receivable as an independent self-contained computer application. Because of these interfaces, many users and software companies integrate accounts receivable processing into comprehensive interrelated systems.

One of the principal advantages of computerizing commercial accounts receivable is often the resulting automaticity of sharing the same input data between diverse applications, and producing as by-products output data required by other applications. By doing this, the amount of manual input preparation and therefore the number of input errors are significantly reduced. A typical commercial accounts receivable system is illustrated in Figure 5.2.

OUTPUT REPORTS

In addition to normal housekeeping reports (error and transaction listings), there are basically three kinds of output reports produced by all accounts receivable systems:

Printed invoices, statements, and other printed notices to be sent to customers

Printed reports for internal use

Computer-readable data required by other computer applications

Invoices and Statements. One of the first things to decide is whether or not to utilize turnaround documents. If the company has more than 100 customers, the invoices and statements can be designed as turnaround

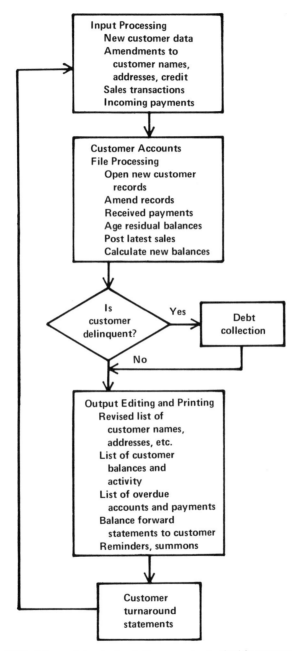

FIG. 5.1   A typical retail accounts receivable system

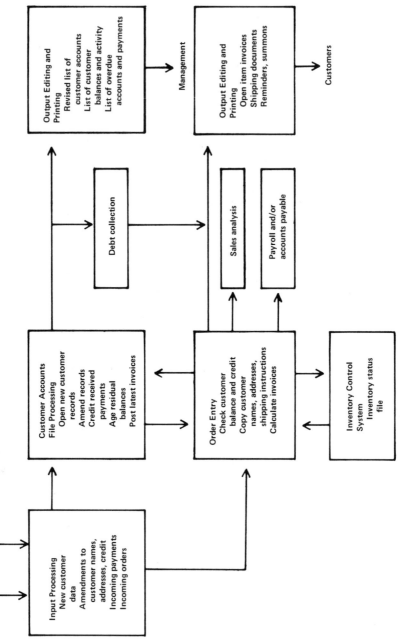

FIG. 5.2   A typical commercial accounts receivable system

74

documents to be entered into the computer system after the customer has returned them with his payment. Naturally, the greater the number of customers, the more likely it will be that the company will utilize turnaround documents. The document should contain all of the data required to be entered back into the system, including the amount. If the customer pays a different amount, this must be noted manually on the document.

If punched cards are used as turnaround documents, a special invoice or statement output program will be required to produce printed data for the customer and punched data for re-input into the computer system. If a special on-line "card printing punch" is used to print and punch the invoices or statements at the same time, the two data sets will be sent simultaneously to the two output buffers on the output device. If an ordinary on-line card punch is used, the two data sets will be punched into two consecutive cards. The cards bearing the punched data will then be processed off-line by an interpreter that will print the data on the cards.

In all cases, invoices, statements, and other documents sent to the customer must be clear, accurate, and attractive. Customers often get a good or bad impression of the company based largely on these documents. Do not try to crowd in too much information. However, accounts receivable system users often differ on which customer invoice or statement layout is most visually pleasing while still supplying all the required information. Large companies may have a forms analyst able to design the required documents. A smaller company might consider the possibility of utilizing the services of a forms analyst from another company or on a part-time "moonlighting" basis. Inasmuch as this is a "one-shot" task, the added cost of hiring an expert in this line will be insignificant compared with the favorable impact that attractive statements will have on customers.

Since all types of variable data must be printed exactly within the spaces on preprinted business statements, the billing/printing programs should provide extremely accurate line counts. Careful document design and completely accurate line count and formatting becomes even more essential if the invoices or statements are to be printed with an optical mark recognition and entered into an OCR document reader when the customer returns them with his payment.

The computer system cannot, on its own, check that the data sent to a printer or punch buffer has been correctly printed or punched into the output document. If these operations are done off-line, it is possible to ensure against such output unit errors by a combination of computer and off-line checks. During the printing (and punching) of each batch of customer invoices and statements, the computer program should accumulate hash totals for the whole batch, then print or punch this total on a control document at the end of the batch. The total of the printed or punched values of all the invoices in the batch is independently accumulated off-line and compared with the total calculated by the computer; the two should be equal.

Printed Reports for Internal Use. These are reports prepared for management, the accounting department, internal auditors, and appropriate supervisors. In contrast with invoices and statements, the printed reports for internal use are not used as turnaround documents. An exception to this would be to design error reports so that corrections can be made on the computer report and submitted to keypunching for re-input into the system. To accomplish this, the error report must be designed as a keypunchable document with card columns shown as part of the report heading.

Some of the other reports usually generated in an accounts receivable system are:

Daily Transaction Register

Accounts Receivable Trial Balance

Accounts Receivable Aging Report

Delinquency Reports and Notices

Customer Name/Address Listing

Customer Mailing Labels

Credit Analysis Reports

Transaction History Reports

The accounts receivable system is often so closely related to sales analysis that these two systems are either combined into a single system or in the minimum have compatible interfaces. In either case, the same customer file should be used.

Examples of other related reports include:

Sales Analysis (by territory, salesman, division, etc.)

Salesmen's Commissions

Sales Tax Reports

Customer Analysis (including activity and payment history reports)

Inactive Customer Lists

Extreme care should be taken in the design of these reports. Some of these reports will replace the 3 by 5 customer payment cards or customer file records of the manual accounts receivable system. Others will be used for management monitoring and control. They must contain all the information on individuals or groups of customer accounts that will be needed. This is also the time to identify all the data that should be included in the customer master file.

Because internal reports must frequently be modified or replaced, it is a good idea to use a report print generator to prepare the programs for them. You can design your own, but most manufacturers provide them with their computers (bundled or unbundled), and it can save time and money to adopt your needs to what is available.

Many of these reports can be summarized for quick review, e.g., the aging reports, delinquency summary, and credit analysis. If the formats of the summary reports are similar to the formats of the supporting detail reports, they will be operationally easier to use. The design of these reports should be completed when you design the detail reports. Now is also the time to design the summary reports that combine and compare data of several detail reports. An example of this would be to take the detailed aging reports and sales analysis reports to combine this data into a management summary report of the open accounts receivable by territory, salesman and division. An aging of these receivables could also be provided.

INTERFACES WITH OTHER COMPUTER APPLICATIONS

All accounts receivable systems generate output data required as input into the general ledger system, whether manual or automated. In addition, the commercial order entry and invoicing procedures produce data required not only by subsequent phases of accounts receivable processing, but also by inventory control, sales analysis, and occasionally by payroll and accounts payable (e.g., for sales commissions). Whether applications are part of the same integrated business management system, or are independent, the output data of each must be in a format suitable for input into each system where the data is used.

It is always easier to program variations in output formats than in input formats. Thus, input/output formats for each of the applications should determine the output formats. In working with the interfaces of an accounts receivable system, the systems analyst must define the data requirements at each interface before he begins designing the system. If the systems are designed by several analysts, the controller must be shown they have jointly defined the requirements at these interfaces and that they are designing based on these requirements.

ESTABLISHING DATA REQUIREMENTS FOR THE MASTER FILES

All accounts receivable systems contain a customer master file. Other files that can be part of the system are accounts receivable detail, transaction, and historical. Often these files are combined into one master file. The design of these files will often depend on the size of your computer and the amount of memory available. If possible, it is efficient to have two files: a customer master file with all current accounts receivable data, and a historical file for old transactions and records.

The following minimum information should be part of the customer master file:

Name of customer

Customer account number

Address to which bills and statements are to be sent

Address to which goods are to be delivered

Credit limit policy of supplier's credit department (the maximum value of goods or services that will be supplied on credit before settlement of the oldest outstanding debts)

Credit and payment terms

Other data that can be stored include receivable balances, date of last activity, salesman data, original amount of transaction, posted amount to transaction, original discount amount, discount taken to date, due date, tax tables, credit/debit memo data, and transaction code.

Historical files are kept when the number of transactions for each customer becomes too voluminous. Old transactions can be purged from the current file once a year. In doing this, the active file does not become too large and can be processed quicker.

In establishing the master files, make sure the analyst allows enough digits to accommodate the largest transaction that can be expected, or, alternatively, that the fields of data are large enough if fields of fixed size are being used. Even with such precautions there will inevitably be exceptions, and rules must be established to modify the various items of data when established bounds are exceeded. Data items especially troublesome are billing and shipping addresses, year-to-date totals, and current invoice amounts.

If the accounts receivable system is receiving data from other related systems, it must provide space to store this information in its own master file:

New accounts data from order entry

Incoming orders and shipping information from order entry

Commissions paid from payroll for subsequent salesmen analysis

If the accounts receivable system is passing on data to other related systems, it must provide space to supply this required information. In most cases, the information is just passed on for storage in the other system's master file but there are occasions, such as credit ratings, where the information is stored in the customer master file (e.g. when an incoming order is received, this stored credit information can also be passed on to the order entry system).

One extremely important field in the accounts receivable system is the customer account number. When automating, continue with the current customer account number if possible. There will be enough problems with conversion.

If new numbers must be designed, construct them in such a way that the digits are related to other data in the file. This ensures that the data is checked automatically. For instance, one digit position or group of digits might be a code indicating the geographical area of the customer's billing address. The computer program can then check that this field in the customer account number and the customer address relate properly to each other; if they do not, an error message on an error report is produced. Another group of digits might relate to the order of magnitude of the credit limit allowed the new customer and his payment terms. In addition, the customer account number can be constructed to contain a check digit equal to the sum of all digits.

PROCESSING RULES

There are seven major areas to be considered when developing the rules of processing.

Establishing New Accounts or Modifying Existing Ones. Customer accounts are opened by recording all the initial file data including an account number in the customer account file. Since the customer account number is usually used to identify all other input transactions, make sure there are control procedures when entering the account number. The controller should get an output listing of all the customer numbers entered to check their accuracy. When embossed credit cards are used in conjunction with automatic reading equipment, the need for manual checking is obviated.

Order Entry and Delivery. Order entry and delivery procedures are required only when suppliers are responsible for delivering their goods or services to the customer's own premises and when an order precedes the receipt of goods or services. Computer users and software houses are sharply divided on whether order entry and delivery procedures are part of accounts receivable processing. In the strictest sense they are not since an order does not become a receivable until the customer gets and accepts the order. This is also the earliest point for recording most retail sales, with order entry and delivery as a separate application. On the other hand, others consider order entry as the first step in a comprehensive accounts receivable system. In either case, order entry must interface with accounts receivable.

One of the first logical steps in order entry processing is the interrogation of the accounts receivable customer master file:

Is the customer placing the order a regular customer?

What is the customer's credit limit and has it been exceeded by previous unpaid orders?

What is the customer's account number required for the input transaction?

If the response is satisfactory, and the order can be processed further, the following typical data is recorded from the customer master file:

Customer's address for delivery of ordered goods

Customer's preferred means of transporting the goods, if any

Customer's address for billing the value of the shipment, which may be the same as or different from the delivery address

The order entry and delivery procedures should then produce a record of the delivered order to the invoicing phase of accounts receivable, after which the order is entered as an unpaid invoice in the customer account file. This interchange of file and transaction data can be greatly facilitated if the order entry and delivery procedures are integrated with invoicing and accounts receivable in the computerized system.

Back Order Processing. When an incoming order is first processed, the order entry system should obtain the following information from the inventory control system:

Are all items ordered in stock?

If so, at what location?

If not, when are the missing items expected?

Missing items must be removed from the original order and a separate holding order must be stored internally in the computer as an automatic input transaction to ship them as soon as the new stock is received. For high-value items that are in stock and can be delivered immediately, the order entry processing will prepare a record of their withdrawal from inventory and the inventory control system will process this record to update the current inventory status file.

Customer Billing. Customer billing is the nub of any accounts receivable system and it is also the procedure with the widest variations between industries, especially between commerical and retail accounts receivable systems.

In commerical "open item" billing, suppliers responsible for delivering ordered goods to their customers normally prepare an invoice for each shipment together or in parallel with the shipping documents. The invoice includes all items listed on the shipping documents as well as their individual prices and the total value of the shipment. In "open item" billing the customer is to pay for each shipment when he receives the invoice. In preparing the invoice, the following operations must be performed:

(1)   Multiplying the unit price of each type of product or service ordered by the quantity to obtain a subtotal
(2)   Adding together the subtotals for all the products and services contained in a single order to obtain a gross total.

In addition to these basic calculations, any combination of the following computations may have to be performed:

(1)   Off-peak or seasonal discounts or price differentials for certain products or services.
(2)   Quantity discounts on a particular product or service item, or in some instances on the whole invoice.
(3)   Special trade discounts applicable to a class of customers or to a particular customer as a result of special negotiations.
(4)   City and/or state sales taxes, which may be chargeable only on some products or only on invoices destined to customers in

certain cities or states, and which may be at different rates for different products and/or different cities or states.

(5)   Freight charges.

In most commercial accounts receivable systems, the invoice will show the payment period and any provisions for early payment discounts. Such information may be preprinted on the invoice, or imprinted by the computer. In retail accounts receivable systems, as well as some of the small commercial systems, "balance forward" billing is frequently used, where the customer pays his bills only on receipt of a balance forward statement sent to him periodically.

Open item and balance forward billing systems have similar characteristics in the sense that many companies will dispatch a balance forward statement to remind the customer that a payment is overdue when they find that certain open item invoices have not been settled within the requested payment period. An open item system can be combined with a balance forward billing because they both occur at different stages of the accounts receivable processing cycle. The calculation of an open item invoice occurs when the goods or services are delivered to the customer and before the value of the invoice is posted on the customer account. But it is only during or after the posting of both the most recent sales transactions and the customer payments that a balance forward statement can be calculated and sent to the customer.

In most retail credit sales, the calculations are performed at the point of sale and the totals accepted by the customer when he signs the invoice before leaving with his goods. Thus, invoice calculations are normally not part of retail accounts receivable processing. In addition, it is desirable in this type of system to free the customer from too frequent payment of small sums of money for individual purchases. Therefore, it is customary to bill a customer only once in each period (usually monthly) for all the purchases he has made since the last bill. Unless there are terminals and a central computer or local minicomputers, retail billing calculations usually consist of adding together the values of a number of individual invoices to produce a statement total. Most retail systems allow customers to pay for purchases over an extended period subject to a certain minimum monthly carrying charge on the outstanding balance. These accounts are often called "revolving accounts." In these cases, "balance forward" billing consists of "aging" the unpaid portion of the previous balance and adding a monthly service charge before adding the value of the current month's purchases.

Processing Customer Payments. In the minimum, input data consists of the customer account number, invoice number, and amount being paid. While this minimizes the amount of manual keypunching, it also increases the risk of the wrong account and amount being credited. The usual errors are simple digit transpositions. Such errors can be minimized by redundant data input checks or source data automation.

Redundant data input checking consists of manually recording additional redundant information data on the input document solely to check on the accuracy of the main identity data. Thus, if the customer is identified on input records both by his numeric account number and the first four or five alphabetic characters of his name, the program can compare both data sets with the corresponding data sets in the customer file. While this type of checking is just another way of making sure the account number is correct, it has the merit of preventing the worst type of input errors in matching payments against the wrong account. Moreover, it is accomplished solely by software and does not require any special hardware. On the other hand, source data automation does require additional hardware equipment.

The validity of the invoice number and payment amount can be checked by the computerized program by referring to the files of issued invoices and statements and confirming:

(1) The invoice bearing the indicated sequence number was issued to the customer with the indicated account number.
(2) The value of the invoice statement with the indicated sequence number is the same as the recorded value of the payment. (This is often difficult to do in that the customer sometimes rounds off payments or accumulates several invoices on one payment).

If the above data are not available, the input will be processed as if it is correct. On the surface, processing of customer payments should be the easiest of all accounts receivable processing, but in actual practice it is generally the most complex. When customers return the stub of their invoice or statement with their remittance, customer account and invoice numbers are available for posting and processing these payments is simple. However, if the customer fails to return the stub, identifying him can be a major undertaking. It has been estimated that an average of 10 percent of retail credit customers and up to 25 percent of commercial customers do not return the stubs or otherwise identify their returns. It is especially difficult when "Bill To" and "Ship To" addresses are different, or when the payee has a name different from the one to which the goods were originally sent.

Even if the customer identifies his return correctly, there may still be the problem (in open item accounting systems) of determining which outstanding invoices are being paid if the sum remitted does not correspond exactly to any of the invoices. Make sure that when computer specifications are being developed, logic rules are included to cover such situations. In many cases, suppliers assume that the incoming payment, in absence of any indication to the contrary, is meant to clear the invoices that have been outstanding for the longest time, but even such a rule as this does not always apply.

Balancing Customer Accounts. All accounts receivable systems, manual or automated, should provide for a periodic (usually monthly) balancing of customer accounts after incoming payments have been credited and the latest credit sales transactions have been charged. Procedures differ, however, according to whether an open item or balance forward accounting policy is pursued.

In open item accounts balancing, a separate record of each invoice is kept on the customer's file until a matching incoming payment has been received from the customer. Since the invoices are supposed to trigger customer payments, most incoming customer payments will refer specifically to one or more outstanding invoices, which can then be cleared from the active customer file and posted to the historical file. Only when an incoming customer payment contains inadequate identifying information will the supplier make the decision to clear the oldest outstanding invoices on the customer's account.

When all incoming payments have been posted and cleared for their respective invoices, and the latest invoices posted to the account, a trial balance, showing the total value of the invoices outstanding on that account, and an aged accounts receivable report should be prepared. At this time "reminder" statements can also be prepared for customers who have exceeded their allowed payment period.

In balance forward accounts balancing, records of individual invoices or sales transactions are kept in the customer account file for only one period after they occur. The most important item on the customer record is the total outstanding balance. From this balance all incoming customer payments are deducted and the residual balance, if any, is aged before monthly interest charges and the latest sales transactions are accumulated to form a new outstanding balance. This new balance is sent to the customer as a part of his new statement.

Debt Collection. When payments for outstanding invoices or statements become overdue, the company may initiate various types of debt collection procedures:

(1)   Mailing reminder statements, in increasingly strong language as the delinquent period lengthens. The invoice and/or balance on which payment is still outstanding may also be included.

(2)   Discontinuing the service, when there is a continuous type of service such as gas, electric or telephone services.

(3)   Taking legal action to recover the money or goods that were delivered but still remain unpaid.

All of these actions may still result in the company not getting paid, and the controller finally has to decide which amounts are to be written off as bad debts.

DATA COLLECTION AND PREPARATION

When transactions occur in high volume, source data recording equipment enabling automatic data collection should be economically justified. In the minimum this equipment will provide a computer-readable input document, e.g., printed from a portable imprinter and the customer's personalized plastic card. For such systems, retail credit customers are issued credit cards embossed with their account numbers which in turn can be read by optical character recognition (OCR) equipment. This account is imprinted automatically on the sales document that the customer signs and becomes a direct computer input document if the value of the transaction is imprinted simultaneously by a point-of-sale transaction register.

In sophisticated systems, it consists of terminals on-line to a central computer — keyboards with printers or cathode ray tube displays, and even minicomputers for calculations at the point of transaction. For such systems, accounts receivable is generally integrated with inventory control, sales analysis, etc., and data entry at the terminal is direct to the master file that services all the applications.

Such systems require additional equipment on-line or off-line to the computer system, and additional costs. The alternatives should be carefully analyzed before the decision is made. Usually the savings achieved by automating the collection of transaction data (e.g., keypunching, costs of correcting errors, etc.) more than justifies the additional cost of the hardware required. However, there are often reasons for this type of automation that are difficult to quantify. These include quicker turnaround of documents and thus quicker processing, better customer reaction with fewer errors, and keeping competitive. Make sure that you consider all of these reasons before you reach a decision.

CONVERSION AND FOLLOW-UP

Before the controller gives his approval to install the new accounts receivable system, the manual accounts receivable records must be converted correctly. An opening balance report of all accounts receivable that have been converted should be checked against the total of the manual records to make sure that all accounts have been converted; the two records must be equal.

Other things to remember before conversion are:

(1)   Training of personnel on new accounts receivable input forms and procedures. (Remember the keypunchers since the manual entries on ledger cards will disappear and input transactions will most likely be keypunched.)

(2)   Training of personnel on the proper handling of new reports, including the correction of error reports.

(3)   Training of accounts receivable personnel on the importance of keeping the customer master file current and correct.

(4)   Notifying the customers of any changes or new invoices or statements.

Parallel Processing. After these things are completed, parallel processing is recommended if there is sufficient manpower available. This double processing should last for at least two weeks or as long as it takes to insure you that the new system is functioning properly. This period should include the end-of-month processing. During this time, make sure that the files are being updated correctly and check to see that customer records on the computerized system are not being lost. This type of control can be gained by a simple one-page daily summary report that lists the opening accounts receivable balance, total of daily transactions, and an ending accounts receivable balance. The total of daily transactions can be controlled in various ways, including batch totals or register controls. If there is a difference in the totals for the two systems, the detailed reports must be checked against all errors and bugs in the new system promptly corrected.

6

Automating Accounts Payable

The following procedures are the basic elements of most accounts payable systems (manual or automated):

Maintaining an up-to-date file of authorized vendors

Validating incoming invoices against fulfilled and accepted deliveries of merchandise or services

Scheduling the payment of validated invoices

Dispatching payment checks

Distributing costs to the department that originated the order for the goods or services

Reconciling canceled checks received from the bank

Until recently it has been traditional for incoming invoices to be validated manually as part of the control over disbursement of the company's cash. Thus, computerized accounts payable systems have tended to be self-contained systems whose processing cycle began with the posting of validated invoices to vendor accounts.

A growing number of companies are now, however, streamlining their cash authorization procedures via the "80:20 rule." This rule allows the controller or another senior financial officer to control only the 20 percent of checks whose face value is above a certain limit (which will vary with the size of the company), since these checks account for at least 80 percent of the company's total cash disbursements. For the remaining 80 percent of low-value checks, payment authorization and invoice validation can be an automatic function performed either by clerks or by a computer.

If this operation is performed by a computer, the computerized accounts payable system will often be integrated with manual or computerized purchasing procedures since both will share at least a common vendor master file. Provisions must be allowed for the recording of new orders placed by authorized departments and the recording of acceptance of deliveries.

A typical accounts payable system is illustrated in Figure 6.1.

OUTPUT REPORTS

Basically, all accounts payable systems produce two types of printed output:

(1)   Reports to the controller and the accounting department at all stages of the processing cycle (in order to retain ultimate control over cash disbursement in human hands).

(2)   Bank checks to be sent to vendors.

Since a printing error on a check could have serious consequences for the company's cash reserves, special controls are normally built into the system. These include:

(1)   An individual listing of all the computer-printed checks with an adding machine tape total. The results are compared with the total value of all printed checks calculated by the computer.

(2)   Computer output of 80-column cards, one for each check; output will contain the check's value in machine-readable code. The cards can then be add-listed by an 80-column card accounting machine and the total compared with the computer calculated total.

(3)   Balancing total value of printed checks plus total value of all updated entries on accounts payable master file, with total value of entries on accounts payable master file at beginning of day's processing plus total value of new orders or invoices.

The format of the bank checks to be sent to the vendor is usually at the user's discretion. The checks should be printed on-line. The checks and the check register should also be programmed to be printed from the same program to make sure that the amounts and check numbers are the same on both the check and the check register. If not, a printing malfunction can cause many checks to be wasted, and when the program is restarted, the check number printed on the check will not agree with the number previously recorded on the check register. This will result in significant check reconciliation problems. To avoid this, the system should contain a restart program that will allow the computer operator to type in the new check number. The program will then insert this new number and sequentially assign the same numbers on both the check and the check register. There is always the possibility of the computer operator typing in the wrong number,

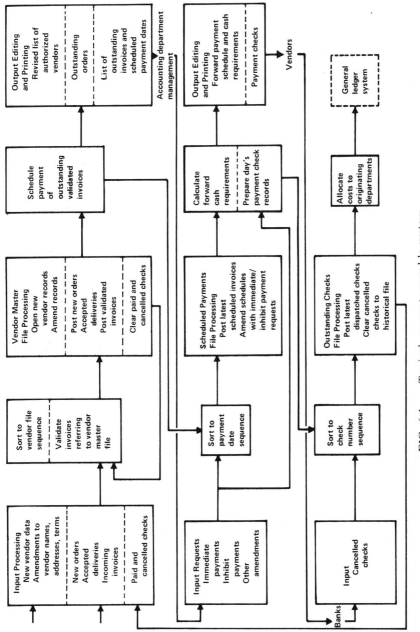

FIG. 6.1 Typical accounts payable system

89

but at least both the check and the register will have the same number. If the restart number results in the assignment of duplicate numbers, the system can be programmed to prevent such an occurrence.

Other reports that the accounts payable system usually produces are:

Invoice registers

Check registers for both computer-produced and hand checks

Credit check lists

Cash requirements by vendor

Outstanding check listings

Canceled check lists

Trial balance

Aged accounts payable

Periodic vendor analysis reports

Expense journals

In most cases, the reports display company name, due date, terms, gross amount, distribution code, invoice number, entry date, department number, invoice date, vendor number, and transaction code. The reports dealing with checks should also show the check number and date paid.

INTERFACE WITH OTHER COMPUTER APPLICATIONS

Most accounts payable systems include a routine for allocating the costs of paid checks to the department originating the purchase order. This cost data is required by the company's cost center or general ledger systems. The purpose of this transfer is to post the required cost data to the appropriate departmental subsidiary ledgers as well as to the company's general ledger. If the company's cost center and general ledger systems are also computerized, the accounts payable system should transmit the required data in the format acceptable by the receiving system in order to eliminate the unnecessary manual data transcription.

Accounts payable systems may also be interfaced with the accounts receivable system to pay certain credit balances on the accounts receivable file. An analysis of the same customers on both the receivable and payable files should be made to allow for possible balancing of these similar accounts.

Another possible interface is with the sales analysis system when commissions payable to salesmen not on the company's payroll must be made. These transactions are usually treated as a check request and can be passed automatically to the computerized accounts payable system.

DATA REQUIREMENTS FOR THE ACCOUNTS PAYABLE MASTER FILES

Types of accounts payable system files include:

(1) Vendor Master File — contains vendor number, name, address, month-to-date and year-to-date payments and discounts, outstanding commitments and discounts, and payment history.

(2) Date Master File — gross commitments to be paid each date, total discount if checks written prior to discount lost date.

(3) Invoice Chain File — vendor number, due date, discount lost date, invoice number, purchase order number, vendor invoice number, part number (if any), batch number, general ledger number, unit of measure, quantity, discount percentage and/or dollars, and invoice gross dollars.

(4) Outstanding Check File — check reference number, check amount, vendor number, vendor name, check date, and voided check data

(5) Master Distribution History File — allocation of expenses for each department or account

(6) Check History File — all checks issued and associated voucher number of each check (by vendor)

(7) Paid Invoice File — details of month-to-date and year-to-date paid invoices

Not all accounts payable systems maintain all of these files but most computerized systems contain the above data in at least a vendor master file, a schedlued payments file, a paid but uncleared check file, and a historical file.

Vendor Master File. The vendor master file is the most important file of the system and must contain the account number and name and address for each active vendor. The file will also have a list of all invoices that have been received and not cleared by the bank, with indicators against each invoice as to whether it has not been scheduled, or scheduled but not paid, or paid and not yet cleared by the bank.

Also included will be the address to which payment checks are to be sent if the address is different from the address where the goods or services are supplied. Many files contain the vendor's usual payment terms, including early payment discounts and service charges, as well as the vendor's current price list for those products and services authorized by the company's purchasing department.

In integrated purchasing and accounts payable systems, the same file will contain a list of all purchasing orders from the time they are placed to the time that an invoice is received for them, with indicators against each order as to whether it is still outstanding, or has been delivered but not yet accepted, or accepted but not yet invoiced by the vendor.

If the vendor master file is used to validate input messages on deliveries of outstanding orders, incoming invoices for accepted deliveries, and inhibit/request payment messages from the accounting department, it will have to be held on a random access auxiliary storage device. This is usually done in most integrated purchasing and accounts payable systems.

On the other hand, the only input data that is validated in most self-contained accounts payable systems is the vendor account number. This can be done just as easily by holding a much shorter file of valid vendor account numbers on a random access auxiliary storage medium or even in main memory, thus permitting the main vendor account file to be held on a cheaper serial auxiliary storage medium like magnetic tape.

Scheduled Payments File. This is normally sufficiently short to be held economically on a random access auxiliary storage device. This simplifies changes or amendments of previously calculated schedules that are subject to modification by the accounting department. Payment records will normally be held in scheduled date sequence.

Paid But Uncleared Check File. Also called the outstanding check file, this will usually be held in check number sequence on either a random or a serial access auxiliary storage device.

PROCESSING RULES

There are seven major areas to be considered when developing the rules of processing.

Authorizing Purchase Orders. Authorizing a new purchase order is part of the procedures of the department that places it, usually the purchasing department, and is not, strictly speaking, part of accounts payable procedures.

However, this is the point when management can begin to control company expenditure in the newer streamlined integrated purchasing and accounts payable systems. The new purchase order must be signed by an authorized individual; in those cases where a budget has been established, the purchase order can be checked against the available budget balance to see if there is sufficient, approved budget dollars to cover the new purchase order. Once the order has been duly authorized and checked for available budget funds, it must be posted to the vendor master file.

The purchase order must have a department number to be charged and an object classification expense code to identify the type of purchase. This data should be included on all purchase orders as it is often difficult to go back and recover this necessary cost distribution information. One way to insure that this date is available and is correct is to check all purchase orders against a valid list of accounts. All orders with errors on department and object classification expense codes would be rejected and shown on an error list which would have to be corrected before the order is accepted. Unless there is a manual, visual check, it is still possible for the computer to accept a valid department number for some department other than the originator's. The only way to catch these errors is to scrutinize the periodic departmental expense reports to make sure that all charges listed were actually authorized by the originator's department.

Accepting Deliveries of Ordered Goods and Services. The criteria for accepting or rejecting delivery of certain orders are again part of the procedures of the department, usually purchasing, that placed the order. In integrated purchasing and accounts payable systems, it is vital that fulfilled and accepted orders be recorded on the vendor master file as that record is important in subsequent invoice validation procedures.

Validating Invoices. This is the first procedure that is normally considered as part of the accounts payable procedures cycle. For those companies that maintain human managerial control over all cash disbursements, invoice validation will usually be performed manually even when subsequent phases of accounts payable processing are automated. In these cases, recording the manually validated invoices against the vendor master file will thus be the first stage of the computerized accounts payable system. On the other hand, in the newer integrated purchasing and accounts payable systems, invoice validation itself may be automated, at least for the 80 percent of incoming invoices whose value is below the limit set by management.

Incoming vendors' invoices are usually validated by checking that the vendor's name and address are listed on the vendor master file for the goods or services that he is invoicing. There must also be validation on authrozied receipt of the goods. This can be done by checking the fulfilled orders file to see that the goods have been satisfactorily received and accepted by the company.

Some companies will also check the unit prices of each item and applicable discounts against the vendors' published price lists, which can be stored on the vendor master file. After this is done, extensions and totals can easily be checked.

After satisfactorily passing all of the company's required tests for validating invoices, the accepted invoice is placed on the accounts payable file with the necessary identification of the department that authorized the order. The type of expense being charged or the object classification code should be included as part of the identification code. This is needed for later detailed analysis of how the various departments and company spent their money. In many cases, budgets are prepared on an object classification code or expense type basis which means that proper identification of all expenses must be made in order to compare actual expenses with the budgeted amounts.

Invoices that do not find a match in the fulfilled orders file will be checked against the outstanding orders file to confirm that they refer to an authorized order that has not yet been delivered or accepted. If a match is found here, those invoices will be placed in a suspense or recycle type file for reexamination during the next processing cycle.

Invoices that do not find a match in both the fulfilled orders and outstanding orders files will be listed on a special report for management action.

Scheduling Invoice Payment. Once an incoming invoice has been validated and placed on the accounts payable file, vendor payment can be scheduled. Most companies seek to minimize such payments by taking advantage of any discount terms offered for early payment, but at the same time not paying any invoice earlier than required in order to maintain their best possible cash position while still maintaining a favorable credit ranking.

The normal procedures to determine the optimum payment date will be to calculate the last date on which maximum discount terms will apply and subtract a set number of days for postal delay. Invoices will then be sorted by their optimum payment date, listed, and totaled to calculate total cash requirements on each payment date in advance. Companies that pay bills only on certain days will schedule the payment of each invoice for the last standard payment day that precedes the optimum payment date.

Some companies have enough buying power to make their own rules for paying incoming invoices irrespective of the vendor's terms, but the above rules will still have to be applied impartially to all invoices when calculating a payment schedule.

Regardless of the principles on which invoice payments are normally scheduled, every accounts payable system, whether manual or automated, must provide for amendments to the routine payment schedule by the controller or other designated accounting personnel. The system must be able to inhibit payment of any invoice until it is authorized, request the immediate payment of an individual invoice for any reason, or lay down revised principles that will be used to reschedule all outstanding invoices.

Paying Invoices. On every payment date, bank checks will be prepared for each invoice that has been scheduled to be paid on that day, and a check register listing these checks will also be prepared. In some companies, all checks and reports must be forwarded to the controller for his signature. In other companies, only the check listing and the checks for amounts exceeding a predetermined limit will be forwarded to the controller; the other checks will be machine-signed and dispatched automatically unless other instructions are received from the controller within a certain time limit. Records for the paid checks must then be transferred from the outstanding list to the paid but uncleared check file.

Allocating Costs. Once the payment checks have been dispatched to the suppliers, the value of each check will be charged internally to the department that authorized the original order. In some companies, this will be done only after the canceled checks have been returned by the bank. The usual practice, however, is to do it when the check is issued.

In those companies that have a cost center system, this payment data is passed on automatically and must include department number and object classification or expense code to be charged. Other details include vendor name, description of the items paid, purchase order number, date of invoice, date paid, amount paid, and any remaining balance.

Reconciling Checks. Dispatching a check to a vendor is not adequate proof of payment. To protect itself from litigation, the company will also require evidence that the vendor has credited the check to his account. This evidence can be provided by the canceled checks. As these arrive, they will be matched against the paid but uncleared check file, which is often an integral part of the vendor master file. Records matching an incoming canceled check will be removed from the file to be added to the final historical file of paid invoices. Many banks offer a check reconciliation service. If your bank provides such a service, make sure that you consider this feature when you are developing your processing rules.

DATA COLLECTION AND PREPARATION

It is vital to maintain proper control over cash disbursements and this means that the accounts payable system must contain various controls for validating manually prepared external inputs to the computer in order to detect and eliminate transcription and other human errors. Here are a number of techniques available for validating accounts payable input.

Establishment of Vendor Numbers with Significant Account Numbers. When a new vendor record is established, all the initial file data has to be recorded manually on computer-readable input documents. Since the vendor master file should be used subsequently to check the validity of other input data, it is essential that no input errors occur at this stage, especially in such numeric fields as the vendor account number, which cannot be easily checked visually.

One programmable checking method is to construct vendor account numbers so that each digit position or group of digits possesses a certain significance related to other data in the file. For instance, one digit position could be a code indicating the geographical area of the vendor's address while another could indicate the vendor's standard payment terms. Since this data should be stored in the vendor master file, it can be checked to prevent incorrect data entering the system. The vendor account number can also be checked by itself if it is constructed to contain a check digit so that the sum of all digits always equals the same number.

Redundant Data Input Checking. In accounts payable processing, redundant data input checking should consist of recording manually additional identification data (e.g., the first four letters of his name) solely to check on the accuracy of the main identity data of vendor account number. While this type of checking is just another way of making sure the vendor number is correct, it has the merit of preventing the worst type of input errors which could trigger a payment to the wrong vendor.

Canceled Checks Input Validation. Input messages referring to canceled checks returned by the bank will be validated normally by reference to the file of uncleared checks. In many instances, however, manual data preparation of canceled check data can be avoided if the bank returns with the canceled checks a magnetic tape listing which can then be read directly into the computer.

In most cases, automated data collection techniques are not recommended for small companies with computerized accounts payable systems. Although the forms are quite varied, they are usually not difficult to keypunch and the volume is not that heavy. When transactions do occur in high volume, it might be that your company has already made use of automatic data collection for other systems and it should be quite easy to utilize this equipment for accounts payable.

CONVERSION AND FOLLOW-UP

After the system has been tested by the analysts, make sure that an independent system test is made by the controller and the accounting users. Immediately before conversion, the manual accounts payable records including information for the vendor master file must be converted correctly. An opening balance report of all converted accounts payable should be checked against the manual total and these two balances must be equal. Make sure that the accounts payable personnel are thoroughly trained to handle their new tasks of batching and controlling the input into the system. Error reports must be corrected, files must be updated properly, and controls must be maintained to ensure that the right vendors are being paid the correct amount.

Parallel Processing. Parallel processing is definitely recommended and should last for at least two weeks. Inasmuch as the manual system is running concurrently with the computerized system, the manual system can continue to be used if there are problems in producing the correct checks. One type of control that can be used is a one-page daily summary of opening accounts payable balance, plus or minus the total of daily transactions, followed by an ending accounts payable balance. Naturally, detailed reports must be utilized to reconcile any of the differences.

7

Automating Payroll

There is one group of operational elements that exists in virtually every payroll system (manual or automated). Basically, these functions compute total earnings and apply deductions to determine each employee's net pay. Specifically, the following quantities are computed:

Gross Pay

Statutory Deductions

Voluntary Deductions

Net Pay

GROSS PAY

Gross pay represents an employee's total earnings prior to any deduction. It is determined using the employee's salary or wage rate and his time, attendance, or job data. In addition to being calculated for the current pay period, gross earnings are accumulated for longer periods such as year-to-date and quarter-to-date. In some companies, an employee may perform several jobs at different wage rates during the same period. In these cases, the payroll system must accumulate these different job earnings and show them as a single total for the employee. However, the individual job details must still be kept as part of the employee's file.

STATUTORY DEDUCTIONS

Statutory deductions are the required federal, state and local taxes that are normally withheld by the company and deducted from each employee's gross pay. For many companies, this capability must include several state and local taxing regulations. In order to properly compute taxes from gross pay, each employee must supply to the company certain tax information. This data, which must be included as part of each employee's file, includes number of dependents, marital status, and taxability. As in the case of gross earnings, the tax withholdings are accumulated for the current pay period, year-to-date and quarter-to-date. However, on certain taxes, such as the Federal Insurance Contribution Act (FICA) Tax, the payroll system must have the capability of withholding portions of an employee's pay only up to a specified cumulative limit.

VOLUNTARY DEDUCTIONS

Almost all payroll systems allow for voluntary deductions. These deductions are amounts contributed by the employee and include insurance (medical, life, total disability), contributions (community fund, pension), and various investments (stock, bond, mutual fund). These voluntary deductions can be withheld up to a specified cumulative limit (limited deductions) or continued until canceled (recurrent deductions). The amount withheld can be either a fixed sum or a percentage of the employee's gross pay. Some payroll systems also include the capability to identify deductions that have been rejected (e.g., those due to insufficient gross pay after taxes), to assign priorities to deduction types to indicate the order in which they are deducted, and to accumulate deductions relative to yearly or quarterly time periods.

NET PAY

Net pay is computed by subtracting the total deductions from the gross pay for each employee. In addition to the usual net pay for the current pay period, figures are also usually kept for year-to-date and quarter-to-date net earnings.

CHECKS AND CONTROLS

In addition to these basic functions, every payroll system should include various types of checks and controls. These checks should be inserted at selected points in the payroll processing to verify the accuracy of the computations. Management must also maintain surveillance over the actual preparation of the payroll to ensure that all pay is computed honestly and accurately. This usually involves comparisons of data pertaining to the entire payroll; e.g., figures representing the number of employees, number of hours worked, and wage rates can be used to determine a preliminary sum of gross pay for all employees. Then, during the actual pay computations, the total gross pay can be accumulated for all employees and compared with the preliminary sum. In this instance, care must be taken to include any payroll adjustments for past payroll periods in these calculations.

The number of employees and number of hours worked can be used to provide further control by comparing the number of persons recorded as having worked to the number actually being processed, and by contrasting the total hours worked to hours paid.

There are many other types of checks and controls that can be included as part of your payroll system. Although the type and number of controls will depend on your individual company's policy, the important thing is to seriously consider these checks and controls as basic functions of any type of payroll system.

The arrangement of these basic functions in a computerized payroll system will naturally vary with each company, but any such arrangement fundamentally consists of the total collection of the included functions, the data flow between them, and the data base from which they obtain information. Using these characteristics, Figure 7.1 is an illustration of a typical computerized payroll system.

OUTPUT REPORTS

Most payroll systems will generate reports for each pay period quarterly, annually, and on request. The usual reports for each pay period include:

Paychecks

Payroll Register

Deductions Taken

Deductions Not Taken

Employee Earnings Record

Payroll or Labor Distribution

Overtime

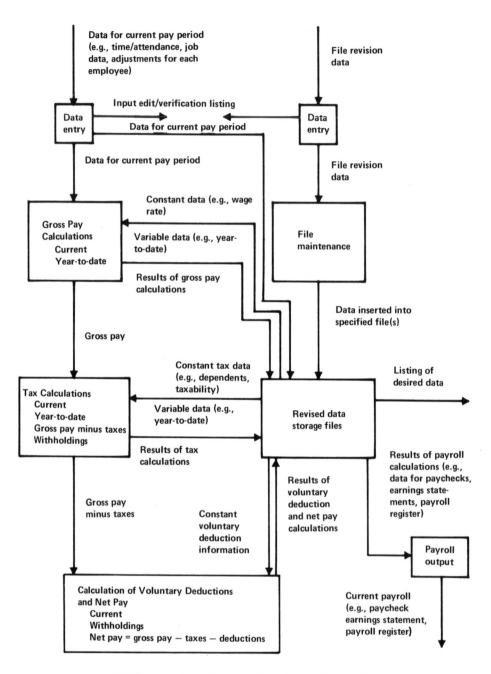

FIG. 7.1   Payroll system functions and data flow

Paycheck Recording Register

Various types of reports, registers, and forms are output to document the results of any payroll processing. However, the paycheck, earnings statement or stub, and the payroll register are outputs of virtually every payroll system. The paycheck is, of course, the vehicle through which the employee is actually paid. The earnings statement or stub, which is usually attached to the paycheck, provides the employee with current pay period and year-to-date data on his pay, taxes, and deductions. The payroll register will usually contain this type of information for each employee of the company. Illustrations of the paycheck/earnings statement and payroll register are shown in Figures 7.2 and 7.3, respectively.

Figure 7.4 illustrates a deduction register and shows the various types of voluntary deductions that can be taken from each employee's pay. There may be different deductions for your company. There may be so many types of deductions that you will want to show only the largest. The others could be grouped into a miscellaneous category. The individual details, however, should be maintained on the employee's master file.

Figure 7.5 illustrates an employee earnings record and shows for each employee the hours worked, base rate, earnings, statutory deductions plus year-to-date earnings and withholding taxes. The employee earnings record is an employee-oriented file containing pertinent employee earnings data over a predetermined time interval. Our illustration covers a one-year period but it can vary.

Figure 7.6 illustrates a paycheck recording register and provides a record of all payroll checks issued by the company. It is often used in reconciling the outstanding checks not yet received from the bank.

In addition to these normal pay period reports, all payroll systems must provide periodic reports required by the government. These are usually prepared on a quarterly or annual basis. Some of these reports, such as the 941-A and W-2 forms, provide the necessary earnings and tax information required by the federal government. Other types of reports may be required by the state or city; others, such as the quarterly tax report which contains various types of earnings and tax information for each employee, may be used by the company internally.

INTERFACE WITH OTHER COMPUTER APPLICATIONS

This system is often very closely related to the personnel system. In many cases, the two systems are combined into one payroll-personnel system. In others, there may not be a personnel system per se and it is the responsibility of the payroll system to supply limited personnel type reports.

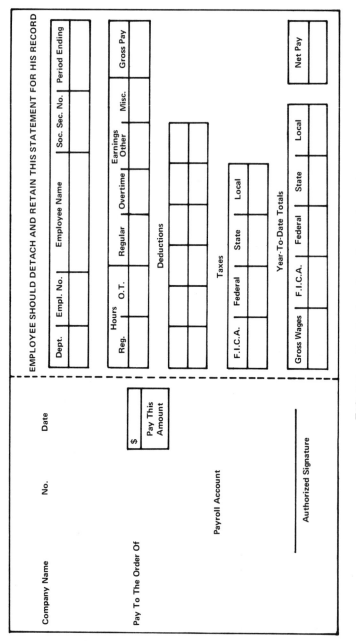

FIG. 7.2   Paycheck and earnings statement

104

PAYROLL REGISTER Date _____

Employee		Name of Employee	Hours Worked	Base Rate	Earnings				Deductions			Net Pay	Year-To-Date		
Dept.	Number				Regular	O.T. Prem.	Other	Total	FICA	With. Tax	Misc.		Earnings	With. Tax	Pay Per.

FIG. 7.3   Payroll register

DEDUCTION REGISTER

Date _____

Type
1-Single Period
2-Specific Period
3-Standing

When Made
1-First Week
2-Second Week
3-Third Week
4-Fourth Week

5-1st and 2nd Week
6-2nd and 4th Week
7-Each Week

| Employee | | Name of Employee | Ded. Code | Type Code | When Made | Savings Fund | Hospital Insurance | Group Life Insurance | Credit Union | Retire Annuity | Total Deductions | Net Pay |
Dept.	Number											

FIG. 7.4   Deduction register

Employee Earnings Record																	
Social Security Record				Name						Address							
Pay Per.	Employee			Hours Worked	Base Rate	Earnings				Deductions			Net Pay	Year-To-Date		Pay Per.
	Dept.	No.	Job Title			Regular	O.T. Prem.	Other	Total	FICA	With. Tax	Misc.		Earnings	With. Tax	

FIG. 7.5 Employee earnings record

Paycheck Register

Pay Date _____

Emp. No.	Employee Name	Check Amt.	Check No.	Emp. No.	Employee Name	Check Amt.	Check No.

FIG. 7.6   Paycheck recording register

In any event, consideration should be given to an immediate or a possible future interface with the personnel functions. Avoid the danger of making the system so complicated that you cannot put any portion of the system into production. It is often safer to plan and design the total system but to install various parts of the system in stages. An example of this is to build a payroll-personnel data base and install the personnel portion first. Get that section working correctly. Be sure that you can add, delete, and change data on this file. When this personnel portion is functioning properly, you have a better chance of installing the payroll portion successfully.

Most payroll systems include a section of their programs for the departmental distribution of payroll expenses. This cost information is usually an integral part of any company's cost center system. Payroll expense represents a significant portion of the company's total expenses and therefore is a vital input for any departmental analysis. If your company does not have a cost center system, the payroll system must generate output data required as input into the general ledger system, whether manual or automated.

Labor accounting is the general term applied to the preparation and output of cumulative labor distribution reports throughout a company. These reports include data on hours on each project, cost per hour associated with each project, shifts worked, and quantities produced. In many manufacturing companies, this type of labor production information is gathered daily and is totaled on a weekly basis for each employee. This data, which includes hours worked, overtime, premium hours, and shift, becomes input for the payroll system so that each employee will get paid for the hours he worked.

Another possible interface is with the sales analysis system when commissions are paid to salesmen who are on the company's payroll. The total dollars paid to each salesman are needed for a proper analysis of the company's sales by territory, salesman, or division.

DATA REQUIREMENTS FOR THE PAYROLL MASTER FILES

Inasmuch as there are many types of payroll systems, there are several types of master files that are maintained in payroll systems. Although many of these are combined, here are the more important ones for you to consider:

(1) Employee Payroll Master File — employee wage records consisting of employee name, address, number, rate or salary, quarter-to-date and year-to-date totals on all earnings, statutory deductions, voluntary deductions, and net pay.

(2) Employee Personnel Master File — personal-type information on each employee, including data on dependents, education, years with the company, positions held, promotions received, military service, special skills, etc. In many companies, this type of

information is combined with the employee payroll master file since there is one common payroll-personnel system. The decision on this will often depend on the role that the Personnel Department has concerning the establishment and coordination of salaries. If they plan a major role, the data on these two files should be combined.

(3) Labor Distribution File — contains data on hours worked, department, shift, overtime, premium hours, absences, idle time, and a detailed breakdown of how each man accounted for his working day. This file is usually a temporary file as the data is summarized and passed on to the employee payroll master file.

(4) Earnings History File — keeps all earnings data by payroll period for each employee which will enable the company to produce a detailed earnings history for each employee.

(5) Detailed Transaction File — records all transactions that are used to update the employee file. This file is usually purged periodically.

(6) Outstanding Check File — check reference number, check amount, check date, and voided check date.

Some of the information stored in these various master files is permanent in that the data does not normally change with each pay period — the personnel data, wage rate, and deduction data groups are in this category — and this information is usually necessary for each pay period's processing. In addition, labor distribution or employee time/attendance data, earnings and to-date data groups contain information that is usually a variable function for each pay period.

In establishing payroll master files, make sure the systems analyst allows enough digits to accommodate the largest transaction expected. Be careful on year-to-date and quarter-to-date fields. Other troublesome fields are names and addresses. Even with such precaution there always seem to be exceptions and rules that must be established to modify the various items of data when established bounds are exceeded.

Another important consideration is the confidentiality of the payroll master file. There should be control keys built into the master file which will only allow the files to be accessed during regular scheduled payroll runs. Along these same lines, it is essential that tight controls be established on the printing and issuance of confidential payroll reports.

PROCESSING RULES

Frequency of Payroll Processing. One of the most fundamental components of a company's payroll policy involves the frequency of paydays. The timing of payroll processing prior to each payday depends upon other decisions, such as the work period included in the paycheck (e.g., period ending in the current week or period ending in the previous week). This is an important policy decision that can greatly affect the formulation of processing rules. If the decision is to pay for the period ending in the current week, rules must be established to handle absences, overtime, dockings, and adjustments for those days that could not be reported in time. At least one to two days are needed to collect the attendance data, keypunch it, run the computer system, and produce and prepare the checks for distribution. If the checks are to be given out on Fridays, it does not allow for any reporting of activity for Friday and possibly even Thursday and Wednesday. For this reason, many companies have a built-in lag in their payroll systems. Company paydays and payroll processing may have periods that are equal but noncoincident. The most common of these periods are weekly, biweekly, semimonthly, and monthly. In a number of cases, companies pay their employees under different schedules (e.g., monthly for executives, weekly for all others).

Due to the confidential nature of the payroll, many companies have a manual payroll system for their top executives. These checks are manually prepared each month and the records for these executives are not maintained on the regular payroll master files. Only authorized personnel have access to these confidential, manual payroll records.

Hiring New Employees. Processing rules must be established to get the required information in the master file on each new employee promptly. Sometimes the paper flow will back up and prevent the new employee from being inserted into the payroll master file, thus he will not get paid by the computer. Manual procedures must be established to handle these cases. Another possibility is that the new input had errors and was rejected by the system. Make sure that the system, in case of new hires, rejects the input only if the necessary data to pay the employee is missing. If miscellaneous personnel data is incorrect, you still have to pay the employee. However, be careful on this point. All data must be correct and complete before it enters the system. Otherwise the information in the master file will not be accurate, and garbage-in, garbage-out (GIGO) will be the results. One way to accept partial data for paying a new employee and still reject the incorrect information is to have a recycle tape as part of the payroll system. All errors appear on this file until corrected and each pay period the errors will again be printed on the error report. This will continue until corrective action is taken to remove the errors from the file.

A further safeguard is to keep track of the number of times an error is recycled and record this number on the error report. A summary report then can be prepared listing those items that have reappeared more than a predetermined number of times. The controller might find this to be an interesting report.

In those companies that report time worked on a daily basis, provisions must be made to accept the new employee's input starting with his first-day activity. If your company uses a recycle tape, these transactions can be recycled until the new hire is placed on the master file. Another possibility is to have the Personnel Department enter the transaction ahead of time, recording a start date. The system would then activate the new man's file on the designated date. However, if the person decides not to take the job, reversing transactions must be submitted.

Calculation of Gross Pay for Various Pay Types. The method used to compute employee wages must be determined in the preliminary formulation of any company's payroll policy. Pay types essentially involve payment of gross pay based upon employee working tme, employee output, and combinations of these two. The basic pay types are:

(1)   Salary — the employee is paid a fixed amount for each pay period. In some payroll systems, salary employees still show an hourly rate in the payroll master file. Thus, if the normal work week is 40 hours and a man makes $100, his hourly rate is shown as $2.50

(2)   Hourly — the employee's pay is based upon the time spent in the employer's place of business. Rules must be established on authorized and unauthorized types of absences. If an employee's hourly rate is $4 and he works only 32 hours, his gross pay will be $128.

(3)   Exception — the gross pay for each employee is calculated in advance, and exceptions to the standard upon which the pay is based necessitate subsequent adjustments. This is the pay type when there is no pay lag in the payroll.

(4)   Piecework — the employee's pay is based upon the amount he produces.

(5)   Incentive — standards are established for each job, usually relative to time required to perform the job or the quantity produced in an hour. These rates must be either entered into the system or be recorded in the employee's master file. Each employee is paid a variable amount in addition to base pay when his performance exceeds the standards. Thus, incentive pay is normally proportional to the amount by which his performance exceeds the standards. If performance is below these standards, the employee still usually receives his base pay. These rules can also be applied to groups of employees as well as to individual workers. Keep the rules simple; otherwise, no one except the worker will know if he is getting paid the correct amount, and in many cases he will only let you know if he did not get paid enough.

Pay Adjustments. These rules can be quite complicated and are often overlooked. Adjustments, which can result in either a pay increase or decrease, are numerous and are utilized by companies in various combinations. When these adjustments affect previous periods and year-to-date totals, they can become quite involved. Backing-out a person's pay can affect tax calculations, amount of deductions to be taken, and distribution of pay expenses. Extreme care should be taken in developing these processing rules. Some of the typical pay adjustments are:

(1) Advances – money paid to an employee prior to the customary time of payment. The advance may reduce the gross pay to such a point that all deductions cannot be taken. Allowances for these exceptions must be included in these rules.

(2) Overtime – money paid to an employee for time worked in excess of the normal work period. There are various overtime categories, such as time-and-a-half and doubletime. Rules must be established to handle all of the established categories. The overtime figures should be kept separately as management often will need details on this type of pay adjustment, not only by individual but by department.

(3) Shift differential – excess employee pay for working a shift other than the first. Rules must be established to cover situations where employees work a split shift or normally work the first shift and are asked to work a few days on another shift. Remember to consider rules for first-shift employees working overtime on second or third shift. Shift differential is sometimes recorded as a fixed amount or a percentage of an employee's base pay.

(4) Bonus – a reward given to an employee, in excess of his normal pay, for a particularly meritorious action (e.g., outstanding performance, extraordinary company loyalty).

(5) Dockings – if the company pays for the period ended in the previous week, this type of docking will normally be reflected in the hours worked. If the company pays for the period ending in the current week, dockings from the previous pay period must be handled as deductions from the current period. In these cases, remember this type of adjustment for any separation pay.

(6) Miscellaneous Deductions – money withheld from an employee's pay to cover a wide variety of items, such as excessive cash advances, purchases from an employee store, and personal telephone bills charged to company facilities.

Establishment of Validity Checks and Controls. Often the controls of a manual system differ greatly from the computerized system. Make sure that you work out in detail with the systems analyst exactly what type of internal and external controls your company requires. One type of internal control is illustrated in Figure 7.7 and shows an example of checks that can easily be included in every payroll system. A sum of gross earnings for all employees is calculated and stored. This computer-calculated figure is then compared with the sum of earnings after taxes and the sum of taxes for all employees. Finally, the sum of taxes, deductions, and net earnings is then compared with gross earnings. Although this type of checking might appear to be unnecessary, it can often highlight bugs in the system that could go

Payroll Processing Phase	Computations for Validity Check	Test
Gross earnings	S_1 = sum of gross earnings for all employees	
Taxes	S_2 = sum of taxes for all employees S_3 = sum of earnings after taxes for all employees $S_4 = S_2 + S_3$	S_1 vs S_4
Deductions	S_5 = sum of deductions for all employees	
Net earnings	S_6 = sum of net earnings for all employees $S_7 = S_2 + S_5 + S_6$	S_1 vs S_7

FIG. 7.7 Typical validity checks

unnoticed. The number and complexity of controls that are established will vary and will depend heavily on the controller's feelings along these lines.

DATA COLLECTION AND PREPARATION

Here is another system that involves cash disbursements. Therefore it is vital that proper controls be established and maintained for validating manually prepared inputs to the computer in order to detect and eliminate all possible transcription and other human errors.

Data to be used in the computer processing runs, such as information on time reporting and payroll adjustments, is entered either directly or indirectly into the computer system. For example, the time card of each employee may itself be used for data entry. On the other hand, data from this time card may be keypunched onto portions of an earnings card which, in turn, will be used for data entry. Automated data collection techniques should definitely be considered in this area as there is a significant amount of fixed data that can be marked on each employee's data card. With the aid of a time clock, start and finish times can easily be recorded. Thus only a minimum amount of variable data has to be manually entered into the system. A cost-justification study should be conducted. Remember to include such factors as quicker processing of the payroll and better employee relations because of having fewer errors on paychecks.

PROPRIETARY SOFTWARE PACKAGES

There are probably more software packages on payroll systems than for any other computer system. If your company is small and does not have a complicated payroll policy, make sure that you explore the possibilities of the software packages. Don't be embarrassed to find out after the fact about a software payroll package that could have satisfied your objectives at a much reduced cost and a significant saving in time. Even if your payroll policy is fairly complicated, it may be a lot easier to modify a few of your standard practices to match an existing software payroll package than to reinvent the total system. Whatever your decision, give yourself time to examine the possibilities.

CONVERSION AND FOLLOW-UP

This is probably the one system for which you cannot be too careful in taking extra precautionary steps to prevent a disastrous conversion. Here are a few suggestions:

(1)   Get actively involved in systems testing.

(2)   Install your system in stages (install the monthly portion first before you attempt the weekly portion).

(3)   Thoroughly train your payroll personnel in new procedures including the proper handling of error correction and control figures.

(4)   Notify the employees of the new conversion and the possibility of errors.

(5)   When the files are converted, make sure that all data has been transferred correctly (this takes time but is well worth the effort).

(6)   Run in parallel for at least two payroll periods making sure that all payroll checks tie in with those produced under the manual system. All differences must be reconciled. All detected bugs should be promptly corrected.

8

Automating Cost Center

Most cost center systems (manual or automated) contain the following basic elements:

ₗA detailed and summarized record of all expenses for each department, project, or cost center

Predetermined detailed budget figures for each department, project, or cost center

A comparison of actual expenses versus budget for each type of recorded expense or budget

Any variances between actual and budget

Each department must be assigned a number, and each type of expense that is to be budgeted must be given its own number. Naturally there are many variations and extensions that you can have in your numbering structure. Some companies have a chart of accounts, and these department and expense numbers are an integral part of this total set of numbers which include codes for balance sheet and profit and loss items, divisions and buildings. For our purposes, however, only department and expense codes will be considered.

DEPARTMENT CODES

Depending on the size of the company, a two- or three-digit number should be large enough to identify each individual department. Departments can be grouped by functions (e.g., accounting, marketing, manufacturing) and numbers assigned accordingly. If there are many departments, a responsibility code can be assigned internally so that the figures for one department can be

117

included in summary reports for more than one grouping (e.g., accounting, division x, vice president–finance, president). This feature will include the department for each responsible group and will eliminate the necessity of re-running the system to get reports for each of the responsible areas.

EXPENSE CODES

In the preparation of most budgets, expenses can be broken down into three main categories:

Salaries and Wages

Supplies and Services

Capital Expenditures

Within each of these major categories, a further breakdown can be made to distinguish the various types of salaries (e.g., managerial, supervision, clerical, direct labor, indirect labor, overtime, premium pay, shift), and a three-digit number could be assigned to designate this specific type of expense. For instance, salaries and wages could be designated as 100s; supplies and services as 200s and 300s; and capital expenditures as 400s and 500s. Thus, the first digit of the three-digit expense or object classification code would indicate the type of major category and the last two digits would provide a finer breakdown of these expenses.

All types of expenses would have to be coded accordingly and any transactions that contained incorrect expense codes would either have to be shown as an error or accumulated in a miscellaneous expense category. The latter is definitely not recommended because the tendency is to leave this category alone and not try to correct it. A better way is to reject the original transaction when it is first submitted to a computer system. Inasmuch as the cost center system receives most of its input from the payroll and accounts payable systems, these two systems would have the responsibility to make sure that only transactions with valid department numbers and object classification or expense codes were accepted. These two systems would still pay the employee and pay the vendor but the transactions with incorrect distribution codes could be recycled on a recycle error tape until they were corrected. Make sure that one person is given this responsibility of keeping these errors corrected. This same person could also have the responsibility of making sure that the department number and object classification code tables were being updated and maintained on a current and accurate basis.

In addition to payroll and accounts payable transactions, the accounting department is usually responsible for preparing the necessary adjustments, transfer of expenses, or journal entries. All of these transactions must also have the correct department number and expense or object classification codes. Remember to include a sufficient description for each transaction so that the department manager will be able to understand the reason for the adjustment when he gets his cost center report.

One more type of input required is budget data, usually inputted once a year. There are two basic types of budgets — fixed and variable. Fixed budgets are mainly used for nonmanufacturing departments whereas a number of manufacturing departments find it difficult if not impossible to forecast a year ahead of time what their expenses will be. Therefore, a manufacturing department will have a fluctuating budget based on the size of the labor force and the amount of standard hours earned. Thus they will not be penalized for peaks and valleys in production levels. Fixed budgets are submitted ahead of time, usually for a one-year period. Details for each type of expense are usually required.

OUTPUT REPORTS

In addition to normal housekeeping reports (error and transaction listings), there are three basic types of cost center reports:

Departmental Summary of Expenses

Detailed Listing of Expenses

Responsibility Summary

Departmental Summary. Figure 8.1 is an illustration of a departmental summary report. In this sample there is a summary line for each type of expense incurred or budgeted. The headings include object classification code, object code description, charges or expenses incurred for the current month, year-to-date expenses, budget, and amount of variance. There can be many variations. Some controllers may want to see percentages or comparable figures from the preceding year. Another possibility is budgeting on a monthly basis and showing budget versus actual comparisons for the current period, quarter-to-date, and year-to-date.

ACCOUNT NO. 124   DEMONSTRATION ACCOUNT   2 MONTHS ENDING 02-28-XX

OBJECT CODE	OBJECT CODE DESCRIPTION	CURRENT MONTH CHARGES	YEAR-TO-DATE CHARGES	ANNUAL BUDGET	VARIANCE
102	Administrative Salaries	2,375.50	4,751.00	28,500.00	23,749.00
103	Clerical Salaries	1,073.80	2,177.80	13,250.00	11,072.20
106	Staff Salaries	4,885.00	9,770.00	58,600.00	48,830.00
127	Overtime	188.50	188.50	–0–	188.50–
***	Total Salaries and Wages	8,522.80	16,887.30	100,350.00	83,462.70*
201	Stationery and Supplies	105.46	255.46	1,000.00	744.54
210	Expendable Equipment	–0–	–0–	700.00	700.00
224	General Supplies	–0–	1,795.00	4,600.00	2,805.00
245	Travel and Entertainment	104.94	104.94	1,200.00	1,095.06
251	Rent Equipment	72.54	122.54	1,000.00	877.46
284	Film	–0–	60.00	300.00	240.00
***	Total General Supplies	282.94	2,337.94	8,800.00	6,462.06*
861	Books	–0–	34.54	–0–	34.54–*
***	Total Capital Expenditures	–0–	34.54	–0–	34.54–*
901	WIP Building Repair	231.80	8,940.35	10,000.00	1,059.65
914	Metered Mail Transfer In	32.46	71.20	400.00	328.80
920	Clerical Stores Transfer	12.50	52.50	300.00	247.50
***	Total Transfer Charges	276.76	9,064.05	10,700.00	1,635.95*
	Account Total	9,082.50	28,323.83	119,850.00	91,526.17

FIG. 8.1   Departmental summary report

Detailed Listing of Expenses. The next type of report shows the details for each type of expense incurred during the current period. For instance, if one department had twenty people on its payroll, the report should give details of what each person was paid for that period of reporting. If the department purchased six items of supplies, there should be details for each purchase. The detailed report should supply the department manager with an accurate record of all of his expenses. Make sure that the report includes sufficient descriptions and corresponding purchase order or identifying numbers which will enable the department manager to reconcile his own records with those shown on the report. In most cases, the department will keep copies of all their placed orders and this cost center detail report can then be used to make sure that the department is being charged properly.

Due to the confidential nature of a company's payroll, the detailed expense report is often separated into a payroll report (Figure 8.2) and an all other expenses report (Figure 8.3). The payroll report enables the department manager to quickly review the salaries to make sure that his department is being charged only for the people who actually worked in his department. An example of this is shown on Figure 8.2 under the overtime category. F. J. Scott is listed under overtime but not under regular salaries. The accounting department could be notified and a quick check of the records would indicate which department should get this charge. Overtime is a type of expense that can build up quickly unless properly controlled. This type of report will enable you to compare actual overtime expenses with the budgeted amounts.

The report showing all other expenses can be given to a clerk who will have the task of reconciling this report to any records kept by the department. Any differences should be promptly reported to the accounting department who should have the responsibility of making all adjustments. An example of this type of error is when one department is charged for an expense that they did not incur. This could have been caused by a transposition or a keypunching error in recording the department number. The accounting department would have to check the original document and see which department should be charged. A journal entry would then be prepared, charging the correct department, and reversing the charge for the department that was charged incorrectly. These details would appear on the next cost center reports as a positive charge for one department and a negative reversing charge for the other department.

Responsibility Summary. The last type of report is a responsibility summary and is illustrated in Figure 8.4. The purpose of this report is to give various levels of management a quick summary of how their areas are doing. For example, if the vice president of finance was responsible for five departments, his report would show a one-line summary for each of his five departments. A quick comparison could then be made on the amount of positive or negative variance for each of his five areas, and if corrective action

| REPORT NO. CC-2 | | PAYROLL EXPENSE DETAIL | | | PAGE NO. 117 |
| ACCOUNT NO. 124 | | DEMONSTRATION ACCOUNT | | | 2 MONTHS ENDING 02-28-XX |
OBJECT CODE	OBJECT CODE DESCRIPTION	EMPLOYEE NAME	SOCIAL SECURITY NO.	EMPLOYEE NUMBER	PAID THIS MONTH
102	Administrative Salaries	R. K. Hillman		02134	1,400.00
102	Administrative Salaries	B. M. Costello		11397	975.50
103	Clerical Salaries	C. L. Thornburg	164 34 2915		604.00
103	Clerical Salaries	C. L. Thornburg	164 34 2915		30.20–
103	Clerical Salaries	W. A. Ryan	191 30 8664		500.00
106	Staff Salaries	D. T. Smith		25868	1,050.00
106	Staff Salaries	G. A. Wilkerson		09144	840.00
106	Staff Salaries	A. E. Hill		13768	775.00
106	Staff Salaries	R. F. Jacoby		00163	1,220.00
106	Staff Salaries	E. P. Parker		26987	1,000.00
127	Overtime	B. J. Scott	177 51 3478		188.50
	Account Total				8,522.80

FIG. 8.2   Payroll expense detail report

OBJECT CODE	OBJECT CODE DESCRIPTION	VENDOR NAME	ITEM DESCRIPTION	PURCHASE ORDER NO.	REQUISITION NUMBER	ENTRY DATE	CURRENT MONTH CHARGES
201	Stationery & Supplies	Lamb Brothers Stat	Inv. No. 14467	CO8751		01-09-XX	17.36
201	Stationery & Supplies	A. B. Dick Co.	Inv. No. 22768	CO6756		01-30-XX	88.10
245	Travel & Entertainment	Holiday Inn	W. B. Harris			02-05-XX	104.94
251	Rent Equipment	Intl Bus Machine	Inv. No. 83915	CO6784		01-30-XX	76.50
251	Rent Equipment	Intl Bus Machine	DB Mem 23561			01-30-XX	50.50-
251	Rent Equipment	Xerox Corp	Inv. No. 66635	C30677		02-24-XX	46.04
***	Total General Supplies						282.94*
901	WIP Building Repair		Work in Process-Aug			01-25-XX	231.80
914	Metered Mail Transfer In		Metered Mail-Aug			02-22-XX	32.46
920	Clerical Stores Transfer		Req. No. 37518			01-24-XX	4.75
920	Clerical Stores Transfer		Req. No. 36104			01-13-XX	7.75
***	Total Transfer Charges						276.76*
	Account Total						559.70

FIG. 8.3   Supplies and services detail report

RESPONSIBILITY SUMMARY FOR TWO MONTHS ENDING FEB. 28, 19XX

NAME	CURRENT MONTH CHARGES	YEAR-TO-DATE CHARGES	CALENDARIZED ANNUAL BUDGET	VARIANCE
Comptroller	$20,000.00	$ 39,000.00	$ 40,000.00	$1,000.00
Budgets	2,000.00	4,000.00	4,000.00	–
Insurance & Taxes	1,000.00	2,500.00	2,000.00	500.00–
Data Processing	40,000.00	81,000.00	80,000.00	1,000.00
Financial Planning	1,000.00	2,000.00	12,000.00	–
Total—V.P. Finance	$64,000.00	$128,500.00	$128,00.00	$ 500.00–

FIG. 8.4   Responsibility summary report

was necessary, the responsible department manager could be contacted. A simple explanation could be given, perhaps an expense being incurred earlier than originally planned or heavier expenses in an area. Often steps can be taken to rectify the situation if these facts are known ahead of time.

The president could get a similar type of report. His report, however, would show a one-line summary for each of his vice presidents including a one-line summary of the total report received by the vice president of finance. Depending on the size of the company, this type of pyramid reporting can be extended to several levels of management.

These basic reports can naturally be expanded into several variations of comparative historical reports. The details supplied by these reports should be quite useful in providing historical data for preparing the following year's budget. Although the flow of information for a cost center system is relatively straightforward, make sure that the reports are accurate, timely, and usable. Figure 8.5 is an illustration of the flow of information of a typical fixed-budget cost center system.

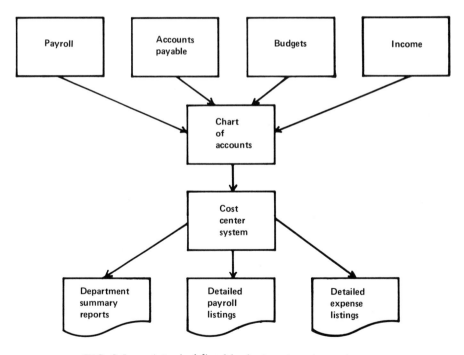

FIG. 8.5   A typical fixed-budget cost center system

INTERFACES WITH OTHER COMPUTER APPLICATIONS

The two main sources of input to this system are payroll and accounts payable expenses. If these systems are computerized, the necessary details can be supplied automatically. This means that if the payroll and accounts payable systems have their last computer runs on the last day of the month, the cost center system has most of its input immediately available. The only other regular input is supplied by the accounting department and these adjustments and journal entries can be scheduled so that the cost center system should be able to be run by the second or third day of each month.

The cost center system is often very closely related with the general ledger system. In fact, if the company does not have a cost center system, this type of information and reporting are usually supplied by the general ledger system, manual or automated. If it is manual, the reports are usually not as detailed, and do not come out as quickly or as often.

This system is also very closely related with the budgeting function. In some companies there is a budget preparation and analysis system. The cost center details will be used as input for this system in order to show historical details of past spending. This printed data will then be used to aid the users in preparing their budget for the following year.

DATA REQUIREMENTS FOR THE MASTER FILES

All cost center systems contain a cost center master file which contains summary figures of all of the expenses for each department. Details of all expenses are usually not kept on this file. If desired, the details can be stored on a history tape and kept for a selected period of time. Remember that you have the printed detailed reports for each accounting period.

Another file that has to be considered part of this system is the chart of accounts file or department number and expense code file. If the list of valid codes is small enough, it can be stored as part of the program, but the real disadvantage with this method is the difficulty of updating the list or table of codes.

The cost center master file must contain all of the required data necessary for the printing of the departmental and responsibility summary reports. These fields include object classification code, object code description, quarter-to-date, year-to-date, current month, and budget details for each type of expense or budget incurred for a particular department. Not all departments will use all of the various expense categories; it is only necessary to keep records on those expenses that are being used.

If the budget is calendarized (monthly or quarterly budgets are prepared), make sure that fields are provided for each month's activity. This information can then be used to show a month-by-month summary of expenses incurred, another valuable tool in preparing calendarized budgets.

As you can observe, the size and complexity of the file will depend on cost center and budget details that your company requires. The issue of how to control the departments that go over their budgets must be answered in the early stages. The areas of budget policy and how the budget is to be prepared must also be adequately covered. Otherwise this task of designing the master file might be a wasted effort and the entire cost center system will have to be redesigned.

PROCESSING RULES

For any cost center system to be truly effective for departmental and accounting use, the reports must be accurate, timely, and readable. This goal should be uppermost as the processing rules are developed. Here are some of the rules to be considered:

Balancing Summary and Detailed Figures. On the surface, this seems to be quite obvious. However, make sure that the totals shown on the detailed reports agree with the totals shown on the summary reports. In addition, the total payroll shown by the cost center reports (plus or minus any accounting adjustments) should equal the total of the checks issued by the payroll system. This is often a good control figure with which to insure the accuracy of the cost center reports. The same type of balancing should also be done with the accounts payable system.

Developing Responsibility Codes. These codes will often be closely related with the company's organizational chart. One department might be responsible to the controller, who in turn reports to the vice president of finance, who reports to the president. Responsibility codes would be assigned to each department for every level of responsibility. Thus, whenever a report summarizing all of the departments or areas directly or indirectly responsible to one individual would be required, it would be very simple for the computer to pick out those departments that had this designated code. Inasmuch as these codes do not change frequently, it is much easier to have this information internally stored. It is not necessary to have this code as part of the department identification number, and enables the system to keep the account number small and easy to use.

Transferring Expenses. Departments within a company often perform services for other departments. Examples of this include data processing, maintenance, supplies, and duplicating services. Caution must be exercised when these expenses are incurred and transferred to the requesting department to make sure that the expenses are not counted twice. In addition, the figures must be grouped in such a manner that a true analysis of the expenses for each department can be made.

One way to accomplish this is to assign object classification codes in the 900 series for any of these areas. Expenses for the requesting department from any of these service areas would be assigned numbers from 900 to 949. The corresponding charges for these transfers of expenses would be assigned numbers from 950 to 999 and could be used only by the service departments. For example, the duplicating service department could be given the object code number 900 to be used for charging all departments using their services. If one department used $25 of these services for one month, this expense would be shown on the requesting department's report under object classification code 900. All of the transfer of expenses of the duplicating service department would be shown under object classification code 950 (always 50 more than the corresponding transfer charge). Thus, the report for the duplicating service department would continue to show all of its original expenses such as labor, materials, machinery, etc., from object classification codes 100 to 699. This total would then represent the cost of operating this department. Below this total, object classification code 950 would show the total of expenses transferred to other departments. An analysis of these two totals could be quite revealing, especially if the costs of these services are compared with those from the outside.

If data processing services are charged out, their numbers could be 901 for other areas and 951 for the total transferred to other departments. Each servicing department that transferred expenses would thus be assigned two 900 numbers, one for charging the requesting department (it could be another servicing department such as duplicating services charging the data processing department) and the 950 - 999 number being used to collect the total of all charges being transferred out of the servicing department.

Thus, an analysis of all object classification codes from 100 to 899 will show all original expenses. The expenses for codes 900 - 949 will provide a total of all transfers of expenses into various departments, whereas codes 950 - 999 will show all of the out-transfer of expenses. Each category will be separate and should enable the controller to analyze his expenses easier and more efficiently.

Scheduling On-Time Delivery of Reports. Included in this category is the determination of the frequency of running. In some manufacturing firms using variable budgets, there is a need to run the cost center system on a weekly basis. In most cases, monthly reports are adequate, but it is vital that these reports are delivered on time. If they are not, much of their effectiveness is lost. The information from the other systems should be available for quick processing but it does not help too much if the accounting department does not have its adjustments ready on time. In addition, after reports have been run on the computer, the carbons must be removed and burst. They are then sent to the accounting department for review to make sure that the figures balance before a copy is sent to each department for review. All of this takes time, and many days can easily be lost if a proper schedule is not maintained.

DATA COLLECTION AND PREPARATION

Inasmuch as the principal sources of input to this system are from the payroll and accounts payable systems, there are minimum problems in this area.

CONVERSION AND FOLLOW-UP

In many cases there will not be a manual system, or if there is, the chances are good that the new computerized system will be significantly different, making the task of running parallel very difficult and possibly a waste of time. Therefore, detailed system testing is a must. The cost center system lends itself very well to semilive testing, which means running a simulated monthly run at the end of two or three weeks with live data. Results can be checked against live production runs and there should be ample time to correct any detected bugs.

The cost center system should be installed at the beginning of a fiscal year, but if it is discovered that the bugs cannot be corrected in time, you could save the input tapes from the other systems for one or two months or you could make provision in the system to adjust the necessary quarter-to-date and year-to-date fields. This type of heavy manual input, however, could result in many errors and should be avoided, if at all possible.

9

Automating Inventory Control

The core of any inventory control system consists of quantitatively reporting for each item of inventory:

When to replenish (reorder point)

How much should be replenished (economic order quantity or EOQ)

How often to review inventory status (review period)

How much stock to maintain in reserve (safety stock)

How is the inventory to be valued (cost)

In practical application, the specific quantity of any particular item may be needed at any time, whereas the cost of the inventory may be needed only at the end of the month or end of the year for financial statement purposes. This current status of all items included in the company's inventory involves the knowledge of and control over inventory additions, deletions, and content, three facets that comprise the general area of inventory control. The basic objective of inventory control is to maintain a level of stock that, within the framework of company policy, achieves the optimum balance between stock depletion and the attendant inability to satisfy customer demands, stock acquisition, and stock retention. For example, the ideal situations in company marketing (enough stock to always satisfy customer demands), manufacturing (long manufacturing cycles), and finance (low inventory cost) may be and probably are incompatible. The task then is to determine the most favorable balance and this is where a good inventory control system can play a vital role.

Fundamentally, a demand for stock is received and the requested items are withdrawn from inventory. At appropriate times, replacements for these items are ordered, acquired, and added to the inventory. Thus, the inventory serves as a buffer between supply and demand. This means that the level of inventory must be set so that the desired balance is maintained. Such planning would be impossible if each item were acquired in response to a specific order. However, the extent to which acquisition can be divorced from consumption is limited because it often becomes prohibitively expensive to carry large quantities of stock for prolonged periods.

Different companies control their inventories using methods that encompass varying degrees of sophistication and all of these methods require large amounts of data. The information, much of which is interrelated, must be accessed, used in the proper manner, and retained. Its utilization frequently involves lengthy mathematical computations, various groupings that reveal specific information (inventory status for ABC groupings of the inventory), and completion of relatively complex forms (e.g., customer orders). The speed, accuracy, storage facilities, and overall data-processing capabilities of a computer system can be extremely helpful in accomplishing these inventory control tasks.

An insight into the wide range of capabilities that exist in inventory control systems can be obtained from Figures 9.1 and 9.2. Figure 9.1 illustrates an inventory management and control system, which is automated to a relatively high degree. It can perform control in both the present and future time frames. In this case an integral part of the operational inventory control procedure is computerized. This system can handle all types of demands from both local and remote sources. It possesses sophisticated forecasting capability and explodes orders to the lowest subassembly level. For the current time interval, and during each of a predetermined number of future periods, demands for an item from the various sources are accumulated. Each demand is accompanied by an actual or projected withdrawal from inventory. At appropriate times throughout the period under consideration, these demands automatically reduce the stock level and calculate the reorder point. If the inventory level is below this computed quantity, an order for additional stock is automatically generated. This order, for an amount determined by the EOQ calculation, is released by the system to the correct production facility at the proper point in time. Subsequently, the progress of work associated with this order is monitored and, when complete, the stock level of the item is increased by the appropriate amount. Moreover, this system includes extensive storage files and sophisticated methods to maintain and interrogate these files.

In other inventory control systems, such as the status and order processing system shown in Figure 9.2, a customer order is entered automatically by either local or remote methods. This demand initiates a system-controlled inventory check to ensure that the order can be filled. If the check fails, a warning message is produced by the system. Otherwise, an intracompany order is automatically generated at the appropriate locations

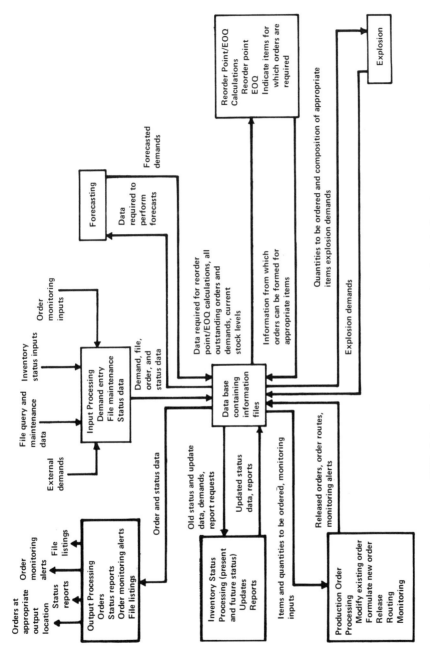

FIG. 9.1   Inventory management and control system

133

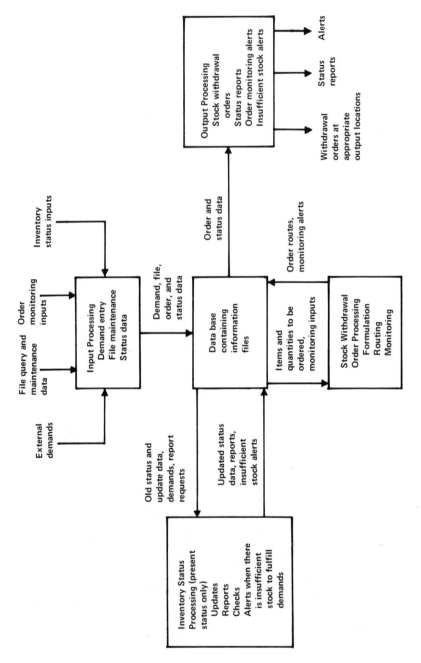

FIG. 9.2   Inventory status and order processing system

134

(e.g., finance department or warehouse). These systems include the automated recording of stock additions and deletions, and the attendant inventory updates. They also include moderately extensive data files as well as file maintenance and query capabilities. The remainder of the inventory control is performed manually. Use of the computer is limited in this illustrated system.

The systems described in the previous examples involve automation of certain facets of the operational inventory control policy. In other systems the preceding computer applications are used primarily as a research tool that provides guidance in formulating this policy, and are not direct priticipants in the actual control of stock. For example, previous demand patterns are statistically analyzed, and the most appropriate method of forecasting future demands is determined. Subsequently, a simulation procedure is initiated in which demands are forecasted and, using these demands, stock level, reorder point, EOQ or other mathematical techniques (such as regression analysis) to determine how much are calculated. This simulation is continued for a specified number of time intervals, and the results are subsequently output to the user. The simulation reveals the effect of varying inventory control parameters and yields a projected profile of the inventory data and associated details during this period.

OUTPUT REPORTS

There are many types of inventory reports that can be produced from an inventory control system. Here are some of the reports and when they should be produced:

Daily

Stock status

Items for manufacturing or reorder

Detailed list of processed transactions

Backorder report

Out-of-stock report

Errors

Weekly

Warehouse replenishment

Warehouse stock status

Open-order status

Monthly

Inventory listing of ROP, EOQ and SS

Out-of-stock summary

Excess or no-usage report

ABC summary of stock status, demand and forecast usage

Summary and details of inventory valuation

Yearly

Listing of old and new rates

Inventory valuation using both the old and new rates

No matter how many of these reports are prepared, the level of each stock item included in the company's inventory must be updated frequently (daily, in many cases) to reflect withdrawals, additions, and adjustments. The status of each item in the inventory is maintained, periodically checked, and reported in almost every type of inventory system. This type of status report consists of some or all of the following:

(1)   General status details (present). Information that reveals the current overall status of all or portions of the inventory is compiled and presented frequently. This information can include item description, stock levels, outstanding orders, production data, forecasted usage, and cost information associated with each item in the inventory.

(2)   General status details (future). The information in this category is analogous to the data described above except that the information is based upon projected future requirements and resultant inventory levels.

(3)   Response to future demands. Upon receipt of a demand for an item that will be required at some future time, the stock level for the item at this future time is checked. This is accomplished using the current inventory and anticipated future stock withdrawals and additions. These demands are used primarily for future planning, and since they do not represent actual customer requirements, inventory withdrawals are not initiated. Because forecasts should be responsive to changing demand patterns, many techniques include provisions for trend or seasonal fluctuations. This responsiveness, however, is often accompanied by a susceptibility to bogus demand fluctuations that decreases the reliability of the forecasted demands. Therefore, since the tolerable level of error in these forecasts is a central ingredient of inventory planning, many inventory control systems include formulas that measure this type of error.

INTERFACES WITH OTHER COMPUTER APPLICATIONS

Inventory control is very closely associated with the order entry function, and in many cases these two systems are combined. At the very least, the same inventory product file should be used. In more sophisticated systems, order entry, inventory control, sales analysis, and accounts receivable can all be combined into one complete marketing system. Thus there would be just one product master file and one customer master file for all of these functions. When an order was received, it would be checked against the product master file to see if the inventory was available. If it were available, the credit rating on the customer master file would be checked, shipping address obtained, and then a shipping paper could be issued and the corresponding inventory reduced. At the same time, an accounts receivable could be established.

If there were not any inventory, an order for manufacturing would be issued and a future order recorded. Even if the systems were not combined into one marketing system, the functions must be interrelated.

Other systems that are closely related to inventory control are inventory costing and work-in-process. Whenever the inventory is costed, these details must be supplied to the inventory system. If the decision is to change only once a year, interfaces must be maintained to transfer this data and to reflect the changes that these new rates will have on inventory evaluation.

Manufacturing companies often record all those finished items that have been manufactured and are ready to be transferred to the finished goods category. Some companies build up an inventory of semifinished parts which are carried as part of finished goods inventory. When a demand arises for a product requiring the semifinished part as the basic raw material, there must be a transfer form the finished goods inventory to the work-in-process category.

Raw materials system is another interface. In some systems, forecasted demand of the finished product can be used to project raw material requirements for as much as one year in advance. These forecasts are of course modified as actual demands change.

Another interface involves the monthly valuation of inventory that is needed for tie-ins with the company's general ledger system. Often given are summaries of the main inventory categories which include the inventory dollar figures for the opening balance, plus additions or packings, less shippings, plus or minus adjustments to arrive at a closing balance.

DATA REQUIREMENTS FOR THE MASTER FILES

All inventory control systems contain an inventory product master file. Another file that is sometimes maintained is a detailed transaction historical file which can often be used to help reconcile differences between the reported balance and any required inventory cycle counting.

Some of the fields of information that are included in the inventory product master file are: description, inventory identification number, current balance, future orders, unfilled orders, number of orders month-to-date and year-to-date, date of last activity, largest order, smallest order, number of customers, largest customer, reorder point, economic order quantity, safety stock, average usage, expected forecast, etc. If the company keeps inventory at several warehouses, similar type information can also be kept for these locations.

PROCESSING RULES

One of the first requirements in establishing any inventory policy is to carefully determine the structure of the various inventory items that your company carries. The inventory items must be studied to determine common characteristics and interrelationships that facilitate item grouping. One way to do this is to list all of the inventory items giving the current inventory dollar value and the expected dollar usage value. These inventory items can then be listed in descending sequence with the largest dollar values listed first. (This is a relatively easy-type computer program.) Another factor to consider in this rating might be the average number of orders per month. In many companies, the experience has been that 20 percent of the inventory represents approximately 80 percent of the inventory usage. Therefore, processing rules will often be quite different for the company's slow-moving items. Categories of ABC (some companies assign more than three categories), often based on dollar usage, are then assigned to each inventory item. The inventory can be reviewed periodically to see if any of the items have moved from one category to another.

In addition to this type of ABC analysis, the size, variation, and urgency of customer orders must be examined to determine grouping and control policy. The applicability of these orders to forecasting techniques must also be considered. The risk of total stock depletion, which can result in unfilled orders, must be set at a level that is acceptable to the company. This is where the ABC analysis will prove to be quite helpful since the company will most likely want to establish different processing rules for the A items (high dollar and fast-moving items) and C items (usually low dollar and slow-moving items). A company with three inventory categories will often have three different sets of rules and these must be determined before programming begins.

Factors associated with obtaining items to replenish inventory supplies must be investigated. For example, the suppliers, acquisition times, and general procedures necessary to obtain these items must be identified. Characteristics of the inventory storage facilities, such as capacity and location, size of the inventory items, and space requirements must also be studied. Various costs associated with inventories must be examined. Typical costs are due to ordering stock, carrying inventory, and encountering stock shortages. Other costs are associated with item production, distribution, and control.

Basically, all of these details are needed to intelligently develop a viable inventory control policy. The inventory items to be controlled, types of control to exercise, and the methods of achieving this control must then be determined.

After the company's inventory policy has been established, there are certain basic processing rules that should be considered for each inventory category:

Demand and Usage Analysis. Customer requirements for an item cause demands to be generated, either directly or indirectly. They may emanate directly from the customer (e.g., entered via his on-line facility) or indirectly from the customer (e.g., entered by the user department in response to customer orders). The system must accept these entries and perform any processing necessary to prepare them for subsequent inventory functions. Rules must be established for future and unfilled orders. If the inventory is available and required immediately, the master file must be properly updated, including all of the usage fields.

Another class of demand is generated within the system from forecasts that combine previous usage with most recent demands to project future requirements. These demands are used primarily for future planning and thus do not initiate any inventory withdrawals.

Another basic type of demand results from exploding an inventory replenishment order for an item into its component parts; for example, demand for an assembly implies that the associated subassemblies are also required. The assembly is thus broken down into all of its component subassemblies (inventory explosion), and a demand for each of these items is automatically generated within the system. This inventory explosion process is continued until demands for the lowest level subassemblies have been formulated. These demands are generated internally and trigger inventory withdrawals whenever necessary to satisfy actual customer requirements.

When all these usages are collected, past demands are analyzed utilizing statistical methods to determine their average and variability as well as any trend or seasonal factors that may be present. Variations of moving averages can be used, or a weighted average, commonly called exponential smoothing, is sometimes used. Basically, the formula is:

New Average = (α) Current Month's Usage + $(1 - \alpha)$ Old Average

Alpha (α) can be statistically determined by further analyzing past demands for each inventory item. Alpha is represented as a percentage and enables the formula to apply a weighted average to the current month's usage while applying another percentage to many months' usage. A real advantage to this type of demand analysis is that only the current average has to be kept in file since the new average when calculated will overlay the old average and thus become the current average. This is in contrast to keeping twelve fields for a twelve-month moving average. This type of averaging does not work for all inventory categories, especially those slow-moving items that have minimum usage patterns.

The calculated new average, which is based on past demand, can then be modified by any advanced marketing knowledge for the new few months. Some companies add a plus or minus marketing factor to the average usage to arrive at a new forecasted or expected usage figure. This technique requires an alert and astute marketing group.

Reorder Point (ROP). Certain demands cause items to be withdrawn from inventory. At some point in time, stock must be reordered to replenish the supply of each item in inventory. The point at which the placing of this order is most advantageous is called the reorder point. Normally, it is set individually for each inventory item and can be determined as follows:

(1)   It can recur periodically following a constant time interval.
(2)   It can be defined in terms of a minimum stock level that must be maintained at all times. This implies that, relative to a collection of inventory status review dates, the algebraic sum of on-hand stock and demands must not be less than the reorder point. In some systems, this type of checking is done daily. When the minimum stock is reached, the replenishment process is automatically initiated. The reorder point can be either a constant inventory level or computed as some optimum combination of lead time, review time, stock usage, and safety stock.

Lead Time. Lead time is normally defined as the elapsed time between the decision to order an item and its availability to satisfy customer demands. It includes such evolutions as order preparation and processing, production by the supplier or your own company, packing and transportation, and receipt prior to availability for subsequent issue. Depending upon the nature of the stock item, its lead time can be either a variable or a relatively constant quantity.

Safety Stock. Safety, or reserve, stock is inventory maintained in case actual demands exceed expected demands. The optimum amount of safety stock for a given item in inventory is a function of:

(1) Carrying costs — the costs of maintaining an item in stock.

(2) Stock-out costs — the costs which are likely to be incurred when all stock is depleted and orders cannot be filled.

(3) Demand predictability. The confidence with which demand patterns can be predicted is often involved in the determination of safety stock levels. For example, considering items having forecasted demands, the variation between these demands and actual usage is used since large errors are frequently offset by a correspondingly high safety stock level. Conversely, if there were no errors in these forecasts, a safety stock of forecasted items would be unnecessary.

These basic factors are usually involved in setting the optimum level of customer service (estimate of frequency of stock-out) which is an important ingredient in determining safety stock. With companies using an ABC analysis, there will often be different safety stock rules for a company's fast-moving and slow-moving items.

Economic Order Quantity (EOQ). This quantity, which is the most economic amount to order, is usually directly responsive to the desired balance between carrying costs and ordering or setup costs, which are those costs incurred in acquiring an item for inventory. EOQ is computed individually for each item and is usually calculated on a monthly or quarterly basis. The amount can be either a constant value or a function of such parameters as average usage, ordering or setup cost, inventory cost and cost of carrying inventory, usually expressed in a percentage. In addition to these basic characteristics, various EOQ techniques employ some or all of the following features:

Upper and lower limits on the quantity that can be ordered

Trend and seasonal adjustments

Calculation of an EOQ with a bias (i.e., either carrying or ordering costs that are higher than optimum values normally calculated) that is specified by the user

Inclusion of price advantages (e.g., those associated with large-quantity purchases)

As in the case of the other calculations, usually there will be different EOQ rules based on the ABC inventory classification rules.

Handling Adjustments. This is a very important part of the processing rules and should not be overlooked. Included in these adjustments are items returned by customers, cycle counting adjustments, items transferred to work-in-process, physical inventory adjustment, and items scrapped. All these items must be accurately recorded and reflected on both the inventory status and the inventory valuation reports.

DATA COLLECTION AND PREPARATION

If there are computer interfaces with the work-in-process and order entry/shipping systems for daily manufactured items and shipments, there will be very little manually prepared input except for inventory adjustments. Controls should be established between these systems to make sure that the additions or packings and shipments are the same as those shown in the originating system.

Redundant data input checking can be used to make sure that all transactions are recorded to the correct account. This type of checking should consist of recording manually additional identification data (e.g., the first five letters of the inventory item description) solely to check on the accuracy of the main identity data of the inventory identification number. While this type of checking increases the amount of keypunching, it has the merit of preventing all types of transactions including adjustments being made to the wrong account. These incorrect transactions could significantly affect the current inventory status and result in the manufacturing or ordering of a slow-moving item that eventually might have to be scrapped.

CONVERSION AND FOLLOW-UP

The new computerized inventory system will most likely be significantly different from the manual system. Therefore, detailed system testing is a must. Make sure that enough time is allowed for thoroughly checking the new calculations such as ROP, EOQ, SS, and expected usage. This type of testing can be accomplished by using a miniature inventory master file with a reliable sample that represents a meaningful subset of the total inventory contents.

Immediately before conversion, the manual inventory records must be converted correctly. An opening balance report of all converted inventory items should be checked against the manual total, and these two balances must be equal. Make sure that the inventory control personnel are thoroughly trained to handle their new tasks. Error reports must be corrected, files must be updated properly, new inventory calculations checked, and controls maintained to make sure that all transactions are being recorded against the right account.

Parallel processing may prove to be quite difficult, but if at all possible, it should be attempted for at least a week. Make sure that inventory figures for both pieces and dollars are checked.

10

Automating General Ledger

All general ledger systems (manual or automated) contain the following:

All financial transactions posted to a set of accounting books, usually a general ledger and subsidiary ledger accounts.

The required data to produce trial balances, profit and loss statements, balance sheets, and other financial type reports.

If the system is automated, the posting of all transactions to the general and subsidiary ledger accounts to which they belong is automatic. This is usually accomplished on the basis of a single input record for each transaction or summary. In addition, if the other financial systems (accounts receivable, accounts payable, payroll, inventory control, sales) are also automated and suitable data set interfaces have been designed between these financial systems and the general ledger system, the single output records for these multiple postings can be done automatically as a secondary operation of the other systems. Computerization of the general ledger system also permits the profit and loss statement, balance sheet, and other financial reports to be produced automatically from the data base. By automating the general ledger system, the reports should be produced many days quicker and more frequently than with a manual system.

OUTPUT REPORTS

At the close of each accounting period (sometimes monthly but always at the end of a fiscal year), the entries in the various ledger accounts are summarized in a number of reports to management designed to give it a meaningful picture of the financial condition of the company. Most systems produce these reports:

Trial Balance. At the end of each accounting period a trial balance of the general ledger, which is a listing of all accounts and their balances, is prepared to make sure that the total of all of the debits equals the total of all the credits. This ensures that the mechanics of the recording and posting operations have been carried out accurately. These account balances are then used as a basis for preparing the balance sheet and the profit and loss statement.

Balance Sheet. The balance sheet lists and summarizes all balance sheet accounts. Asset accounts are listed on the left-hand side of the report while liabilities and shareholders' equity and retained earnings (net worth) are shown on the right-hand side or, in some cases, underneath the assets. The assets are listed as current, fixed, and other. The liabilities are usually shown as current and noncurrent.

As a practical matter, the difference between a current asset and other (noncurrent) asset is easily understood. However, the defining of an exact boundary between these two categories is not as easy. Usually there are five types of assets included in the current asset classification:

Cash — money in any form

Secondary cash reserves — marketable investments and securities

Short-term receivables — includes both open accounts receivable and short-term notes receivable

Inventories — raw materials, supplies, work in process, finished goods

Short-term prepayments — cost of services paid in advance

The difficult area is the distinction between short- and long-term investments in productive goods and services. The American Institute of Certified Public Accountants (AICPA) states that the test usually applied in distinguishing current from noncurrent assets is whether the investment of these assets will be realized within one year or the operating cycle of a business, whichever is the longer period of time. The average lapse of time between the investment in materials and services and the final conversion back to cash is usually accepted as the length of the operating cycle of a business.

Fixed assets are the long-term tangible resources used in the operation of the business. Equipment, land, buildings, machines, and tools are included in this category. There is a growing trend, however, to group all assets other than current assets and long-term investments as noncurrent assets. In addition to the fixed assets, there are two other identifiable types of noncurrent assets:

Long-term or restricted funds, investments, and receivables

Long-term intangible resources — goodwill, copyrights, patents, etc.

The difference between current and noncurrent liabilities is not as difficult to define as it was for the assets. Current liabilities are usually defined as obligations that will require the use of current assets to satisfy them within the next year or operating cycle, whichever is longer. A noncurrent liability would naturally include all other liabilities. There are three main types of current liabilities:

Obligations for goods and services — accounts payable, payroll

Short-term notes payable

Advances or deferred revenue — collections received in advance of services or delivery of goods

The shareholders' equity and retained earnings or net worth in a business is the residual interest in assets, after liabilities have been deducted. Naturally, this calculation can result in a plus or a minus net worth.

There are many important indicators found on the company's balance sheet. One of these is what proportion of assets is balanced by liabilities, and what proportion by shareholders' equity and retained earnings; the higher the latter proportion the healthier the business is. Others include earnings and dividends per share; net working capital, which is the amount by which current assets exceed current liabilities; debt-equity ratio, which is the ratio of long-term debt to stockholders' equity; and the quick, or acid-test, ratio, which is obtained by dividing total quick assets by current liabilities. Quick assets are those current assets that can be turned into cash immediately. Inventories, slow receivables, and prepaid expenses are not considered quick assets.

Income and Expenses or Profit and Loss Statement. At one time the balance sheet was considered the primary end-product of accounting. However, this is not the case today as there are many accountants who consider the balance sheet as secondary to the income and expense statement. This income and expense statement, often called a profit and loss statement, lists and summarizes all operating accounts. Income accounts are listed on the left-hand side of the statement while expenses are shown on the right-hand side, or underneath the income accounts. By definition, the sum of operating income and losses equals the sum of operating expenses and surpluses. However, the important indicator is whether there are operating losses or profits and how big a proportion of income is represented by surplus, or how big a proportion of the operating expenses is balanced by losses rather than income. Similar statements are prepared for each subsidiary ledger, thus permitting management to assess which departments, projects, and, in some cases, products make the biggest contribution to the company's total profits and which cause the greatest loss.

For most companies the major source of income is the production and sale of goods and services. This type of information is readily available from most sales systems. Income offsets are usually distinguished from expenses and should be deducted from gross revenues or income as part of the income section of the income and expense statement. Examples of these offsets are sales discounts and sales returns.

Expenses are classified in the income and expense statements by the nature of the expense elements, business functions, areas of responsibility, or any other way that proves to be useful to the company. However, for both expenses and income, remember that a good income and expense statement is something more than just an itemized list of income and expenses. As controller, you should give real consideration to the type of classification (single-step versus multiple-step), the amount of detail needed, the order of presentation, and the titles to be used. All of these points must be agreed upon before output specifications are completed.

Important indicators from the income and expense statements are trends in sales and earnings; return on total assets, which is net income divided by total assets; return on long-term capital, which is net income plus interest on long-term debt divided by total assets minus current liabilities; and earnings and dividends per share. A strong earnings record often means a strong financial condition. Although this is definitely not the whole story, it is often a good indicator that a company with a proven earnings record can usually work out its financial problems.

Budget Analysis and Variance Reports. A straightforward balancing of income and expenses will not be particularly suitable to control the operations of such departments as research and development, whose expenses on specific projects are always incurred months or even years before its projects reach the production, and therefore the income-producing, stage. In an expanding enterprise, especially one involved in a rapidly advancing technology, the current expenses of its research and development departments will normally exceed the current income from the sales of previously developed products, even though these products are profitable in terms of the research and development (R&D) expenditure that was devoted to them.

Thus, the only fair method of controlling this type of expenditure is to plan a budget carefully for each project, in line with marketing forecasts of what can eventually be earned, and then to compare actual expenditures with planned expenditures. This is the function of budget analysis and variance reports in which actual expenses incurred in developing a certain product or in any other type of budgeted activity are compared with the amount of money alloted for it, and the variance from budget targets is computed both in absolute values and in percentage terms.

Needless to say, the same method of control against budgetary targets can be employed for income as well as for expenses incurred by projects or departments in which current income is balanced more closely with current expenses. A project or department may be profitable but not nearly as profitable as management had planned.

Another type of variance report is the cost variance. When future expenses are budgeted, notably in a research or development project, they are usually considered on the basis of a schedule of standard costs for given quantities of the materials and the labor required on the project. (A running check should be kept on the validity of these standard cost assumptions.) The object of these cost variance reports is to compare the actual costs incurred as calculated from payroll timesheets and suppliers' invoices for materials supplied with the standard costs for each identifiable type of labor and material. Any resulting variances can then be computed both in absolute terms and in percentages and may be of real value in explaining the reason for the budget variances pinpointed in budget analysis reports.

In those companies that have cost center systems, these reports can easily be included as part of that system; however, if your company does not have a cost center system, this type of budget analysis and variance report would become part of your general ledger system.

Comparative Historical Reports. Another method of income, expenses, and profitability control is to compare each department's and project's current performance with its performance during earlier comparable periods. Where performance is liable to be affected by seasonal factors, it will be usual to compare current monthly figures with those of the same month in preceding years and current year-to-date figures with those of the equivalent portion of preceding years. When performance is unlikely to be affected by seasonal factors, current monthly figures are often compared only with those of preceding months.

Types of comparative historical reports include:

Comparative balance sheet

Comparative income and expense statements for the company as a whole and for individual projects or departments

Comparative budget analysis and variance reports

Comparative cost variance reports

In each case the variance from performance or balance sheet entries in preceding months or years can be highlighted both in absolute terms and in percentages.

All of these output reports of a general ledger system are internal documents destined to circulate only within the company's accounting department and top management. There should therefore be a standardized fixed-field format for all of these reports especially since the format will be dictated largely by current accounting standards. There will naturally be wide variations in the number and names of different ledger accounts kept by different companies. Practice varies even between accountants regarding the titles to the main reports, notably whether to call the operating report the "statement of income and expenses" or the "profit and loss statement" or a variation of either title, and what title to give to departmental budget analysis reports, cost variance reports, and other reports required by the same system. Whatever titles are used, companies should be consistent so that when reference is made to a particular report, all personnel in the company will know exactly which report is meant.

Computerized general ledger systems can produce a large variety of reports. However, the reports previously discussed are the ones most frequently produced. A typical general ledger system that will produce these reports is illustrated in Figure 10.1.

INTERFACES WITH OTHER COMPUTER APPLICATIONS

The general ledger system is the one system that probably has an interface with every other computer application. Inasmuch as the general ledger is often the last system to be automated, companies should be constantly thinking of the requirements of the general ledger system as they are designing the other computer applications to make sure that the necessary information is available in the system. Otherwise, modifications will have to be made to the original system. In most cases, however, the data is available, since in the absence of a computerized general ledger system, the company has to maintain some form of a manual general ledger system, and if the data supplied by these other systems was insufficient, changes should have been made at that time.

DATA REQUIREMENTS FOR THE MASTER FILE

All computerized general ledger systems will have at least one ledger accounts master file but will differ in whether they include account posting

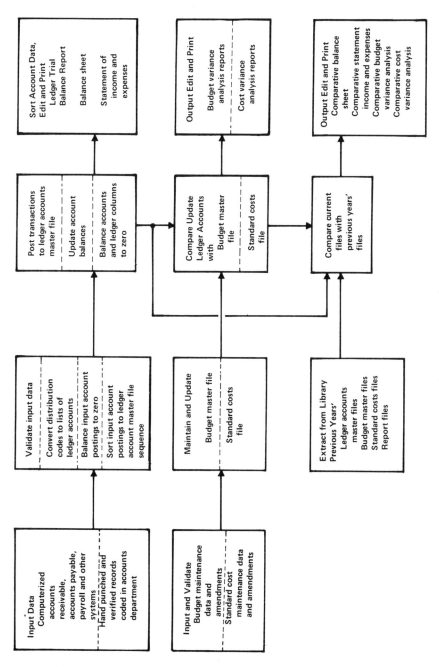

FIG. 10.1   Typical general ledger system

151

distribution tables, departmental budgets, and historical data on this ledger accounts master file or on separate files.

Ledger Accounts Master File. The ledger accounts master file will normally include all general and subsidiary ledger accounts. Unlike manual ledger accounting systems, most computerized ledger systems will tend to hold all accounts of a certain type together and treat ledger distinctions as a lower level of referencing than the type of account. Thus the general ledger sales operating account will tend to be preceded by all subsidiary departmental sales operating accounts. This inversion of the normal off-line classification system is purely one of internal computer system convenience, however, and does not affect the manner in which reports are presented.

Both the posting of transaction summaries to multiple accounts and the summarizing of various accounts for reports can be facilitated if the general ledger accounts master file is held on a random access auxiliary storage medium, but this is by no means essential. Many systems will regard the infrequency of ledger posting runs as adequate justification for holding the ledger accounts master file on the most inexpensive storage medium available – magnetic tape – even at the expense of a larger number of sorting runs both prior to posting transactions and prior to producing reports.

A company's general ledger master file may have a few dozen to several hundred distinct accounts, depending on the complexity of the business, and these in turn may be paralleled by an equal number of accounts in each divisional or departmental subsidiary ledger. These accounts are either balance sheet or income and expense records.

Each balance sheet account – whether on the assets or the liabilities side – is posted with both credit and debit transactions during a normal accounting cycle. Thus the ledger account corresponding to a current bank account will be credited with all the checks from the customers and other debtors that are paid into the relevant bank account and debited with all the checks that the company draws on the bank account to pay its obligations. At the close of each accounting cycle, the final balance recorded in the ledger account for the current bank account involved should equal the closing balance recorded at the same time in the bank's books.

Each balance sheet account record in the general ledger should contain at least the following: ledger account number, description, opening balance at beginning of current accounting cycle, total credits during the current accounting cycle, total debits, and closing balance. Some systems will provide for entries on the opening balance and cumulative credits and debits for two or more distinct accounting cycles, usually a monthly and an annual cycle.

Each operating or income and expense account record in the general or subsidiary ledgers will contain at least the following: ledger account number, description, and total transactions in the current accounting periods to date. Transactions are usually recorded cumulatively for at least two accounting periods simultaneously, the current month and the current fiscal year. But there is nothing to stop any company from having additional entries for weekly or quarterly performance figures. Companies may record the details of individual transactions that occurred during the current accounting cycle, but the most usual procedure is to process these in other computer systems and to record only the sum of all the transactions of a given type that occurred during a given accounting period in the relevant ledger operating account.

Account Posting Distribution Tables. When single short distribution codes are used on transaction records to indicate the list of accounts to which the transactions concerned should be posted, the length of the distribution tables required to translate these codes into a string of account numbers will depend on the complexity of the general ledger accounting system. In the simpler systems, the tables may be short enough to be held entirely within the computer's main memory, in which case they can be held on the program tape or at the beginning of the ledger accounts master file. In more complex systems, it may be necessary to hold them on a separate auxiliary storage file.

Departmental Budgets. The departmental budget data required to compare departmental income and expense performance against budgets will be held on the ledger accounts master file in some systems and on a separate file in other systems. If your company has a cost center system, this information will normally be stored in the cost center master file. Even when this information is held on a separate file, data should be organized in the same sequence as the general ledger accounts master file to facilitate matching of the two files during budget variance analysis runs.

Historical Ledger Data. To produce historical comparative balance sheets and comparative income and expense reports, a general ledger system will require access to general ledger account data that could be one to two years old. This historical data can be retained on the general ledger accounts master file as long as required for historical comparative reports, but the usual procedure is to match the current updated general ledger accounts master file with a historical file containing data for 1, 3, 12, and 24 months previously.

PROCESSING RULES

Ledger Postings of Details. Companies differ in the extent to which they record details of individual transactions in general ledger records. The normal procedure is for detailed transactions to be processed in lower-level accounting systems for accounts receivable, accounts payable, payroll, inventory control, etc., and for the general ledger accounts to be used only for recording the sum of all transactions of a specific type that occurred during a given accounting cycle. In the case of bank accounts, day-by-day accounting for incoming and outgoing checks is left to the bank concerned, with company checking at the end of the accounting cycle and posting of a summary of all credit and debit movements to the ledger balance sheet account.

Frequency of Ledger Postings. When only summaries of transactions are posted to general ledger accounts, posting will rarely occur daily but will most likely be concentrated toward the end of the company's lowest-level accounting cycle. If the shortest accounting cycle is one month, then postings will occur at the end of each month; if it is a week, postings will occur at the end of each week. This does not preclude more frequent posting of general ledger accounts regarded as key indicators, such as bank accounts, cash in cashiers' tills, and operating accounts for sales, labor, and material costs.

Posting to Multiple Ledger Accounts. Regardless of whether individual transactions or only summaries of transactions of a given type are posted to ledger accounts, each transaction or transaction summary will require posting to several different accounts. It will usually affect one general ledger operating account and parallel operating accounts in one or more subsidiary ledgers and will require posting to two or more general ledger balance sheet accounts, which will be paralleled sometimes but not always in certain subsidiary ledgers.

For example, sales of finished goods from inventory control should be credited to the operating account for sales in both the general ledger and the subsidiary ledgers of the division or department responsible for the goods' manufacture and distribution. At the same time, they should be reflected as debits on the balance sheet assets account for inventory and credits on the balance sheet for either cash or accounts receivable depending on the type of sale. If the sale price was higher than the standard cost at which the value of the goods sold was recorded in inventory, this type of transaction affects both the balance sheet and income and expense statement, as it is reflected as part of the company's profit.

Zero Balance Checks. These postings to multiple ledger accounts can be quite complex and general ledger systems must guard carefully against posting errors and omissions. The most usual method is to balance the accounts to zero at the end of each account posting cycle in two different directions. On the one hand, each general subsidiary ledger account is balanced to zero individually on the basis of the following formula:

Opening Balance **plus** Total Credits **minus** Total Debits **minus** Closing Balance

On the other hand, the four vertical columns for opening balances, total credits, total debits, and closing balances are individually balanced to zero for all accounts on the basis of this formula:

Balance Sheet Asset Accounts **plus** Operation Expense Accounts **minus** Balance Sheet Liability Accounts **minus** Operating Income Accounts

Whenever any of the four columns does not balance to zero, errors will be found in the account entry that also does not balance to zero horizontally.

DATA COLLECTION AND PREPARATION

Coding the Single Input Record. One of the major advantages of a computerized general ledger system is that transactions or transaction summaries can be posted automatically to all the general and subsidiary ledger balance sheet and operating accounts to which they belong following input of a single transaction record.

Even this single input record does not necessarily need to be keypunched and verified manually. Each of the other computerized financial systems can automatically pass the necessary information to the general ledger system as long as the necessary linkages have been designed in each of the related systems. Naturally this can reduce manually prepared input for general ledger computer systems to infrequent file maintenance data.

Regardless of how the single input record is prepared, it must contain indicators specifying to the general ledger system all the ledger accounts to which the transaction or transaction summary has to be posted.

Direct Listing of Account Numbers on Input Records. One way of routing the input transactions correctly to their appropriate accounts is to list all the account numbers to which they have to be posted on the input records themselves. This procedure will tend to be followed:

(1)   In small companies, where the general ledger is simple, and each type of transaction requires posting only to a limited number of ledger accounts.

(2)   During general ledger conversion runs.

(3)   For exceptional types of transactions, for which no standard distribution has been foreseen.

In large companies, however, many types of transactions or transaction summaries might require posting to as many as a half dozen general and subsidiary ledger balance sheet and operating accounts, each of which might be identified by an account number up to ten digits long.

The interfaces between the other computer applications and the general ledger system would be greatly complicated if each system had to list a long string of ledger account numbers on each transaction summary sent to the general ledger system. On the other hand, if the coding of input records to the general ledger systems with the appropriate account numbers is left to accounting department clerks, this would require almost as much clerical effort as to post each transaction summary directly to the relevant accounts by hand. Following the clerical coding, the input records would have to be transcribed manually to a computer-readable device; and the longer the list of account numbers on each, the greater the risk of data preparation transcription errors.

Indirect Addressing of Input Records via Distribution Codes. The alternative is to put only a single distribution code on each general ledger input record. The computer system can translate these codes into the list of all the ledger account numbers to which the record has to be posted by reference to a distribution table held in its main memory or on an auxiliary storage device. When additional ledger accounts are created, only the distribution table has to be updated, and, therefore, no changes have to be made to the distribution codes or to the other computer systems.

In some cases, the operating account number may play the role of the distribution code. This is true if the range of operating accounts has been designed in such a manner that all postings to any one operating account always affect the same balance sheet accounts. Meanwhile, the subsidiary ledger operating accounts that are affected and the general ledger operating account can be identified by a combination of operating account number and organization code. This does not entirely eliminate the need for distribution codes or quasi-operating account numbers for those transactions that affect only balance sheet accounts, such as incoming payments from credit customers.

Ledger Input Validation Checks. No matter how the input comes into the general ledger system, manually or via computer applications, coding errors affecting either the general ledger account numbers or the transaction amounts must be carefully avoided. There are two types of validation checks employed in computerized general ledger systems, and these are:

Check digit, batch totals, and file reference validation

Balancing ledger inputs to zero

As in other computer applications, ledger account numbers can be constructed to incorporate a check digit so that the sum of all the account number digits always comes to the same figure. During the original general ledger input run, each account number field can be checked for validity by totaling all its digits.

Input records can be grouped in batches, and all the account number fields of each batch can be accumulated to hash totals, while a batch total is accumulated simultaneously for the transaction values of all the records in the batch. Hash totals and batch totals can then be compared with the totals previously accumulated off-line during manual data preparation or during the output run of the other computer applications. In addition, the validity of the account numbers can be checked against the ledger accounts master file to confirm that each listed account number does indeed exist.

Ledger inputs can be further validated as a whole by balancing all the inputs of a processing cycle to zero. Credits to balance sheet asset accounts and operating expense accounts will be added to debits from balance sheet liability accounts and operating income accounts. All debits from balance sheet asset accounts and operating expense accounts and credits to balance sheet liability accounts and operating income accounts are then deducted from the previous sum. The total must balance to zero.

CONVERSION AND FOLLOW-UP

This is one system that should be very easy to run in parallel. System testing can be limited as it is more effective to run the full system in parallel with live and complete data. Make sure that your accounting people are thoroughly trained and prepared for any additional or changed reports. The reports may also come out quicker and more frequently and the office schedule will have to be adjusted accordingly. The accounting system may even be modified, but in all cases your people must be informed.

Index